William Whewell is now recognized as the first systematic writer on the history and philosophy of science. This book shows how he engaged in metascientific criticism in the early Victorian period, when the social and intellectual status of 'science' and 'scientists' were still matters of contention. In coining terms such as anode, cathode, physicist, and scientist, he contributed not only to scientific vocabulary, but also to the language in which science is now discussed. Whewell was concerned with the nature of science in the broadest sense – its history, ethos, metaphysical foundations, religious, educational, and social implications, and the intellectual and moral biography of its practitioners. These issues were discussed in a public forum in books, journals, addresses, and sermons. Whewell was arguably the major, but not the sole, participant in these debates; yet more so than others, such as John Herschel and David Brewster, he sought to justify a role for himself as a metascientific commentator while still establishing the language and concepts of history and philosophy of science. Whewell's work can therefore be seen not only as an attempt to define science, but to clarify his vocation as its leading critic.

IDEAS IN CONTEXT

DEFINING SCIENCE

IDEAS IN CONTEXT

Edited by Quentin Skinner (general editor), Lorraine Daston,
Wolf Lepenies, Richard Rorty and J. B. Schneewind

The books in this series will discuss the emergence of intellectual traditions and of related new disciplines. The procedures, aims and vocabularies that were generated will be set in the context of the alternatives available within the contemporary frameworks of ideas and institutions. Through detailed studies of the evolution of such traditions, and their modification by different audiences, it is hoped that a new picture will form of the development of ideas in their concrete contexts. By this means, artificial distinctions between the history of philosophy, of the various sciences, of society and politics, and of literature may be seen to dissolve.

The series is published with the support of the Exxon Foundation.

A list of books in the series will be found at the end of the volume.

DEFINING SCIENCE

William Whewell, natural knowledge, and public debate in early Victorian Britain

RICHARD YEO

Senior Lecturer, Faculty of Humanities, Griffith University, Queensland

CAMBRIDGE
UNIVERSITY PRESS

Published by the Press Syndicate of the University of Cambridge
The Pitt Building, Trumpington Street, Cambridge CB2 1RP
40 West 20th Street, New York, NY 10011-4211, USA
10 Stamford Road, Oakleigh, Melbourne 3166, Australia

First published 1993

Printed in Great Britain at the University Press, Cambridge

A catalogue record for this book is available from the British Library

Library of Congress cataloguing in publication data
Yeo, Richard, 1948–
Defining science: William Whewell, natural knowledge, and public debate in early
Victorian Britain / Richard Yeo.
p.   cm. – (Ideas in context)
ISBN 0 521 43182 4
1. Science – Great Britain – History – 19th century.   2. Science – Philosophy –
History – 19th century.   3. Science – Historiography.   4. Whewell, William, 1794–1866.
I. Title.   II. Series.
Q127.G4Y46   1993
509.41′09′034 – dc20 92–30620 CIP

ISBN 0 521 43182 4 hardback

For
Mary Louise, Gillian, and Claire

Contents

Preface   *page* xi
List of abbreviations   xiii

PART ONE

1   Introduction   3
2   Science and the public sphere   28
3   Metascience as a vocation   49

PART TWO

4   Reviewing science   77
    (1)   Science and reviews   77
    (2)   Whewell's early reviews   87
5   Moral scientists   116
6   Using history   145
7   Moral science   176

PART THREE

8   Science, education, and society   209
9   The unity of science   231

List of references   256
Index   276

Preface

My interest in the place of science in early Victorian culture began with postgraduate work on natural theology as a philosophical framework for scientific investigation in that period. Whewell was a figure in this thesis and has been in other work I have published. But it is only recently, and with the benefit of some further specialist studies, that it has seemed possible to use Whewell as the focus of a book on debates about the nature of science in nineteenth-century Britain. Inevitably, this brings problems of selection, since nothing like comprehensive attention can be given both to Whewell's activities and to the range of issues in the public discussions in which he participated. Consequently, I have not attempted to include detailed accounts of Whewell's interventions in particular scientific disciplines such as mineralogy, geology, or electrochemistry, or his contributions to terminology. Where possible, I indicate other relevant treatment of these topics. I have aimed to place Whewell's work in its cultural context and therefore hope that this book will interest not only people who already know Whewell, but also students of Victorian intellectual history more generally.

Although this book was written over the last few years, it has a prehistory. During this time, which I'd rather not quantify, I have happily gathered many debts. An inadequate attempt to recount them should start with thanks to the librarians of Fisher Library, Sydney University, and Griffith University Library, especially the interlibrary loans staff who ferry the large Victorian holdings of Australian institutions. The staff of Trinity College Library, Cambridge, have provided vital assistance, and I am also grateful to the Master and Fellows of the College for permission to use papers in their custody and to reproduce the portrait of Whewell on the jacket. I also thank the staff of the British Library, the Library of the Royal Society, London, the National Library of Scotland, and Edinburgh

University Library for their assistance and access to manuscript collections. The Faculty of Humanities, Griffith University, has supported the project over several years, and I have benefited from a university research grant and study leave. The final writing was done at the Institute for the Advanced Studies in Humanities, University of Edinburgh, where I was a Visiting Fellow in 1992. I would like to acknowledge that some of the material in chapter 9 appeared in my contribution to the collective volume on Whewell published by Oxford University Press in 1991.

Sometimes less tangible, but always crucial, have been the advice and interest of scholars and friends. John Brooke, Geoffrey Cantor, Menachem Fisch, Jonathan Hodge, David Knight, Rachel Laudan, Don McNally, and Simon Schaffer have shared enthusiasms for the problem of 'dealing with Whewell', and I value their conversation over the years. John Gascoigne and Dorinda Outram offered constructive comments from their respective comparative perspectives, and kindly read drafts of some chapters. I also thank David Oldroyd for giving me his spare copy of the *History*. Errors of fact and other shortcomings are my own.

Without the support, empathy, and sense of humour of my family, none of these other encouragements would have prevailed. I dedicate the book to them.

Brisbane
June 1992

Abbreviations

BL   British Library

HP   J. F. Herschel Papers, Library of the Royal Society of London

NLS   National Library of Scotland

RS   Royal Society

WP   William Whewell Papers, Trinity College Library, Cambridge (small superior figures in WP references refer to folio numbers allocated to individual letters and pages thereof)

PART ONE

CHAPTER I

Introduction

In the early Victorian period there was a wide-ranging set of debates on the nature of science. These embraced topics such as the ethos, method, epistemology, and religious and social implications of natural science, the moral and intellectual character of its practitioners, and the historical development of its theories and procedures. We recognize some of these topics today as parts of the philosophy of science, of the history of science, or of science policy. Some of them are now confined to the domain of specialized and professional scholarly disciplines; others intersect with political and social controversy in the public realm.

William Whewell (1794–1866), the Master of Trinity College, Cambridge, wrote two monumental works on the history and the philosophy of science before these became specialist and technical subjects. He did this at a time when the social and intellectual status of 'science' and 'scientists' were still matters of contention. This is partly reflected by the fact that Whewell coined terms such as anode, cathode, physicist, and scientist, thus contributing not only to scientific vocabulary, but to the language in which science is now discussed. Apart from his major works, Whewell engaged with other writers in a discourse about the nature of science in reviews, addresses, and sermons, and in doing so established himself as the leading critic of science – a role that perhaps compares with the cultural criticism of contemporaries such as Coleridge and Carlyle.

It is difficult to imagine how the early Victorian debates on science could be analysed without confronting Whewell's presence. Similarly, an historical understanding of what Whewell was doing requires some reference to the purpose and scope of his project, the ways in which it was pursued, and its reception by contemporaries. Until recently, however, most studies of Whewell's philosophy of science paid little attention to the institutional and social context in which he lived and worked, or to the intellectual and cultural status of the phenomenon he analysed – the scientific enterprise. It is interesting that this strongly 'internalist' approach resisted challenge much longer in the case of ideas on the nature of science than it did where substantive scientific theories were concerned. One result is that we now have a more rounded understanding of Charles Darwin's work than we have of Whewell's, and not simply because the former is a more influential figure. This difference also indicates the strength of twentieth-century professional demarcations which abstracted philosophy of science both from the actual practice of science and from other philosophical and intellectual concerns.

This in turn reflects the divorce between science and consideration of values since the late nineteenth century. When Max Weber lectured on the scientific enterprise in 1919 he spoke of vocation, personality, and values – three concepts that seemed, by this time, to be removed from scientific thought and activity (Weber 1989). Weber was writing as science became more closely involved with the State, thus raising the question of whether value judgements had to be removed from science so that the State alone could set the goals of research and determine its application. In the 1930s, the philosopher Edmund Husserl regarded this development as part of 'the crisis of European sciences' – science, although successful at one level, no longer had anything to say about human ideals and aspirations (Husserl 1970, 5–7).[1] Whewell was concerned to explain the success of science, but he did this within a discourse that had not yet become separated from discussion of other cultural issues, partly because science had not yet become firmly dependent on the State. With his contemporaries, he had to consider not only the relationship between science and values; he had also to affirm the value of science.

Since the demise of extreme positivism Whewell's work has gained more respect. It no longer figures simply as the convenient target for the winning message of J. S. Mill's *System of logic* (1843) and its attack on idealist epistemology. This is not to suggest that all his views have found acceptance among contemporary philosophers of science; on the contrary, his work is still marginal to the mainstream. Nevertheless, it is admitted that the problem about the relationship between theory and observation – so crucial to debates since the 1930s – was clearly posed and elaborated through historical examples in

---

[1] Husserl located this abdication in the sixteenth century.

Whewell's *History of the inductive sciences* of 1837, and again in *The philosophy of the inductive sciences* of 1840. Since the 1960s, his emphasis on the role of hypotheses has been compared with the position of Karl Popper, his account of dramatic theoretical change with that of Thomas Kuhn (Schipper 1988). But this recovery of Whewell can be taken further, in another way. Recently there has been discussion of the manner in which political values are involved in the different philosophies of science advocated by Popper, Imre Lakatos, and Paul Feyerabend (Ravetz 1984; Chalmers 1991). Given this awareness with respect to modern theories of science, it would be odd if the rediscovery of Whewell abstracted *his* ideas about science from the intellectual and social context in which they were developed. Indeed, his case offers a chance to see how some of the major assumptions about science were debated in a framework that included values – religious, moral, and political. One contention of this book is that Whewell's work needs to be firmly located within this wider context, and not solely in one determined by the priorities of twentieth-century philosophy of science.

William Whewell was one of the leading men of science in the Victorian age. His life spanned the period of transition from natural philosophy to what we now recognize as modern science. Whewell coined the term 'scientist' in 1833 but it was not generally adopted in Britain until the close of the century, partly because some of the important men of science, such as Michael Faraday and T. H. Huxley, preferred to think of their work as part of broader philosophical, theological, and moral concerns. This issue of the relationship between natural science and other cultural values and activities occupied a major part of Whewell's extensive writings. But unlike Faraday, Darwin, Humphry Davy, Louis Pasteur, and James Clerk Maxwell, his name is not linked with any single major scientific discovery, and thus his reputation has not been consolidated in standard histories of science. One of the ironies of Whewell's case is that as the outstanding historian of science in his day he reinforced the perspective in which past science is viewed as a story of heroic individuals – usually great *men* – wresting secrets from Nature. In so far as his name was not firmly linked with such a discovery, Whewell was relegated to the sidelines of the story by his own historiography.

In this respect, the comment of Francis Galton in his *Hereditary genius* of 1869 was prophetic. After noting that reputation in science was heavily influenced by the association of an individual's name with some striking discovery, he suggested that this could lead to the neglect of others whose work contributed to scientific progress. He then cited Whewell as an example of one who, in spite of being amongst the most able of his generation, was destined to be forgotten:

His intellectual energy was prodigious, his writing unceasing, and his conversational powers extraordinary. Also, few will doubt that, although the range of his labours was exceedingly wide and scattered, Science in one form or another was his chief pursuit. His influence on the progress of Science during the early years of his life was, I believe, considerable, but it is impossible to specify the particulars of that influence, or so to justify our opinion that posterity will be likely to pay regard to it. Biographers will seek in vain for important discoveries in Science, with which Dr. Whewell's name may hereafter be identified. (Galton 1892, 186; also Lightfoot 1866, 13–14, on the problem of how posterity would judge him)

This quotation also poses the problem of specifying exactly what Whewell did accomplish, given that it was not substantive scientific discovery. We are faced with the fact that Whewell's status in the scientific community was prominent, and there is strong testimony from contemporaries about his influence on scientific debates, especially in the years from 1830 to 1850. Although he did do some important scientific research, and received a medal from the Royal Society for his research on tides, Whewell's reputation derived largely from his informed and incisive reflections on the work of others and, indeed, on the nature of science itself. Beginning with his early textbooks on mechanics, written while a Fellow of Trinity College, and culminating in his mature role as the magisterial historian and philosopher of science and Master of Trinity College, Whewell created a space for himself in the scientific world as an authoritative commentator on the progress of science, its method, epistemology, and cultural significance. One more well-known illustration of this is the fact that he was approached by both Faraday and Charles Lyell when they were searching for a new terminology in which to express their scientific ideas. In thus becoming a legislator of the language in which science was presented, Whewell occupied a significant and influential position within the scientific community.[2]

Sydney Smith's quip – that science was Whewell's forte but omniscience his foible – has to a large extent produced negative responses to the man and his work. One reason for this is that the goal

[2] For accounts of Whewell's activities, see Cannon 1964; Robson 1964. On his invention of terms, see Ross 1961. Schaffer 1991 puts this in the context of a wider reform of language.

of universal knowledge has become impractical; even in Whewell's time it was beginning to be seen as unreachable. But as well as being astonished or sceptical about the range of Whewell's interests it is important to understand the assumptions behind them. It is then possible to gain some perspective on the intellectual context in which the natural sciences developed in early-nineteenth-century Britain. Whewell's career coincided with the rapid expansion and specialization of science – perhaps a second scientific revolution – and this made the scope of his interests vulnerable to Smith's humour. However, there was another sense in which the ideal of embracing a range of sciences was supported by contemporaries, not so much because they believed it possible for one man to keep abreast of *all* developments in several sciences, but because the general overview it offered was regarded as valuable. Leading men of science such as Charles Lyell, Adam Sedgwick, and John Herschel, although they did not agree with all of his views, applauded Whewell's work in history and philosophy as a contribution to an understanding of how science developed and how it might proceed. Nevertheless, as revealed in the extensive and intimate correspondence with some of his closest friends – especially Richard Jones and Julius Hare – Whewell had to convince himself about the value of this critical reflection.

The difficulty of finding a single label for Whewell's intellectual achievements has affected existing scholarship. Repeating twentieth-century divisions, this has compartmentalized his work, treating him as a philosopher, an educationist, a natural theologian, or a scientific manager. The recent collection of essays edited by Menachem Fisch and Simon Schaffer portrays the involvement of Whewell in these and other spheres, but the editors indicate that these interrelated in ways that are difficult to specify (Fisch and Schaffer 1991, vii–viii). Philosophers of science such as Buchdahl, Butts, Laudan, and Ruse have provided helpful analyses of Whewell's major writings on science, and their publications and those of others in this field remain the largest part of Whewell scholarship. However, this approach has been more concerned with the relevance of Whewell's thought to particular issues in the philosophy of science – such as induction, theory change, the role of hypotheses and epistemology – rather than with the relationship of these ideas to his other interests. Given this, the importance of Fisch's recent study is that it details the emergence of the intellectual problems Whewell sought to solve, and shows how his mature views took shape (Fisch 1991a). In contrast, this book does not treat these technical, philosophical issues in detail; rather, it examines the cultural context in which Whewell's project emerged, the different forums in which he pursued it, and the implications it contained for discussions of education, moral science, the relation between science and technology, and natural theology.

In spite of this renewal of interest in Whewell, however, there is no large-scale study of his work and life, no detailed intellectual biography. Isaac Todhunter's account (1876), a muted version of the usual Victorian lives and letters, has not been superseded in the way that similar productions devoted to Faraday, Lyell, Darwin, and Kelvin have been. One reason for this is clear: the extraordinary polymathic range of his activities, from mineralogy to moral philosophy, has been a daunting obstacle. But in another sense it may be that the nature of Whewell's intellectual career does not fit the current genre of scientific biography. Here we may see a vindication of Galton's prediction, namely that study of Whewell has been conditioned by his failure to conform with the model of a scientific life structured around significant discoveries.

If individual reputations were secured by 'discovery', we need to ask why Whewell felt that science itself needed careful justification. Until we have an appropriate category for understanding the intellectual enterprise sustained by this perception, it is difficult to see how any comprehensive biography of Whewell would be possible. I suggest that this activity might be seen as metascientific commentary or criticism, realizing that this concept is susceptible to the danger of anachronism already confronted by philosophical reconstructions of his ideas. Nevertheless, Whewell's case seems to call for explicit recognition of the fact that he created for himself a role as the critic and reviewer, adjudicator and legislator of science. Moreover, he did this at a time when the scale and character of the scientific enterprise was changing in fundamental ways. These changes laid the framework, both in terms of institutions and values, for what we now recognize as 'science'. Thus one of the caveats about the use of the term 'metascientist' is that Whewell was not simply observing and distilling a stable phenomenon; rather he was seeking to influence the way in which science, its method, epistemology, and values, was defined and promoted. Whewell was an actor, as well as a critic.

The aim of this book is to highlight the *activity* in which Whewell was engaged by analysing his metascientific commentaries and critiques, paying attention to the various forms in which they were made. It is fairly well known that Whewell was one of the most argumentative of the Victorians. It is less often noticed that he exploited the full range of media available to nineteenth-century authors – books, reviews, pamphlets, addresses, lectures, and sermons – and the result was that his methodological and epistemological reflections on science were associated with theological, moral, biographical, and historical preoccupations. These were the elements often sidestepped by some later commentators in order to reconstruct Whewell's contribution to philosophy of science. My contention here is that Whewell's theoretical views on science were formulated and justified within an attempt to defend and affirm its moral, intellectual, and cultural value.

As well as creating a vocation for himself as a metascientific critic, Whewell influenced the form in which such analysis was made. When he began to think about the nature of science, the philosophy of science did not exist as a specialized discipline. By the time he died it was beginning to crystallize as a distinct discourse. Whewell's own extraordinary two-volume *Philosophy* of 1840 contributed to this process, but it derived from a much broader set of cultural concerns. Thus one of the aims of the chapters which follow will be to consider the different circumstances and media in which Whewell made his various commentaries on the nature of science. He wrote periodical reviews, history, philosophy of science, natural theology, moral philosophy, and entered existing debates on biography and education. In some cases he was able to shift the agenda; in others he was constrained by existing forms and assumptions. With this in mind it is possible to appreciate the various contexts in which he was involved and the tensions which are present in his work. Some of these stem from the paradox that the man who coined the term 'scientist' increasingly pursued an intellectual project in which the word 'science' almost returned to its former meaning – systematic knowledge of any subject, including ethics, architecture, language, and politics.

INTRODUCING WHEWELL'S PROJECT

Rather than attempting a concise summary of Whewell's main arguments on the philosophy of science, I want to offer a picture of what he sought to accomplish in his major works, and how he explained his aims to readers. One way to achieve this is to sample one of the forms of writing in which Whewell excelled: prefaces (or introductions).[3]

We can begin with the opening pages of the *Philosophy*, since these included references to the *History*. In the preface and introduction (rewritten for the second edition) Whewell explained his aims and made claims about the novelty of his project. He stressed that it was a contribution to the 'philosophy of knowledge', not just the philosophy of the physical sciences. The latter were the best starting-point because they offered examples of universally recognized and stable truths (Whewell 1840a, i, ix, 3–4). In principle, however, the analysis of these subjects – astronomy, geology, mechanics, chemistry, optics, acoustics, botany, and physiology – was applicable to other fields of inquiry in which 'man's knowledge assumes that exact and substantial character which leads us to term it Science' (1840a, 5, 16). This formulation is doubly significant. First, relatively new disciplines in which general laws had yet to be established were included, and Whewell listed these in an order that upset the usual hierarchy which favoured the mature disciplines. Second, Whewell defined 'science' as a kind of knowledge, rather than as a kind of subject-matter.

Whewell also specified his project by the manner in which he described his predecessors. The most prominent of these was of course Francis Bacon. Whewell saw himself as an inheritor of the programme initiated by the *Novum Organum*, but also its beneficiary: he was now in the position of being able to study the history of scientific progress which Bacon, in his time, could only imagine, and to some extent, prophesy. The comparison with nearer contemporaries was less gracious. Admitting that there were philosophical investigations of knowledge which made reference to physical science, Whewell emphasized that these had not been selected as part of a systematic, historical survey; they were merely 'detached examples' taken from one or two branches of science. In a veiled reference to Scottish philosophers, such as Dugald Stewart, he noted that some of these accounts were based on political economy, philology, morals, or the fine arts – all important subjects, but not yet suitable for this purpose (1840a, 9–10). Significantly, when Whewell named the 'great

---

[3] For introductions to Whewell's philosophy, see Blanché 1967; Butts 1968; Laudan 1971; Losee 1983; Ruse 1976. In this chapter I cite the first edition of Whewell's *Philosophy*, but elsewhere use the second edition of 1847, and the third edition (1857) of the *History*.

masters of the philosophy of science' they were not modern philosophers or metaphysicians: they were Bacon, Isaac Newton, and
Georges Cuvier. Noting that the views of these writers differed from
his, Whewell declared that 'upon maturer consideration' they would
have embraced his doctrines (1840a, xii).

Whewell thus presented himself as doing what no other writer had
done. There were grounds for this claim, but the confidence with
which he announced it in 1840 belied his earlier anxieties about the
best way of explaining his self-appointed task as the definer of
inductive science. However, he did admit one problem, to which we
will return in later chapters: namely, the need to comment on several
specialist areas of science. Alluding to the recent criticism of the
section on physiology in the *History*, he acknowledged a clash of
interests:

Those who have well studied that subject, feel a persuasion, a very natural
and just one, that nothing less than a life professionally devoted to the
science, can entitle a person to decide the still controverted questions
which it involves; and hence they look, with a reasonable jealousy, upon
attempts to discuss such questions, made by a *lay* speculator. (Whewell
1840a, 1, xii)

This is a most revealing expression of the indefinite status of the task
Whewell had chosen.

When he began to outline the central theme of the work, Whewell
admitted another difficulty. Stressing that his purpose was to
'establish distinctions, not to obliterate them', he argued that the
distinction between 'Fact and Theory' was not a simple one. There
was 'a mask of theory over the whole face of nature'. Whewell
approached this not by outlining existing epistemological theories but
rather by appealing to everyday examples in which apparent facts,
such as the motion of the earth, involved theoretical assumptions.
These, he claimed, suggested that 'a fact under one aspect is a theory
under another'. Furthermore, they showed that the mind was active
in perception: 'The scene of nature is a picture without depth of
substance, no less than the scene of art; and in the one case as in the
other, it is the mind which, by an act of its own, discovers that colour
and shape denote distance and solidity' (1840a, 23–4).

In this first edition of the *Philosophy* Whewell did not fully articulate
his doctrine of the fundamental antithesis. He subsequently developed this in 1844 and then remodelled the introduction of the second

edition. However, the elements of this principle were present: the intimate relation between sensations and ideas, objective and subjective sources of knowledge. He contended that all knowledge depended on the practical union of the terms of these, and 'many other antitheses'; equally, philosophy required their analytical separation. Reporting the common view of the mind as a passive receiver of knowledge from the world, Whewell asserted that it was active. Here he introduced the term 'Idea', but said he was using it in a more specific sense than the manner in which it was often 'used for almost all imaginable results of our passive and active powers combined'. Ideas were not simply 'transformed sensations'; they were the active element that gave 'form' to sensations. 'We cannot', he explained, 'see one object without the idea of space; we cannot see two objects without the idea of resemblance.' These and other 'Fundamental Ideas', such as time, number, and cause, supplied 'Ideal Conceptions' appropriate to the various sciences. Whewell then gave an unfamiliar slant to the term 'Induction', claiming that it took place when ideal conceptions allowed the mind to 'connect the facts', to unify previously disconnected observations under a rule or law. Kepler did this when he showed that the planets moved around the sun in elliptical orbits. But he did not do so merely by accumulating observations; instead, after struggling with various conceptions, his mind '*superinduced*' upon the facts the idea of the ellipse (Whewell 1840a, 25–45; see Fisch 1985 for a full account of Whewell on the fundamental antithesis).

By stressing the active role of the mind in induction, Whewell gave this concept a meaning at odds with prevailing interpretations. In contemporary treatises on logic, such as the influential one by Richard Whately, induction was an inference governed by syllogistic form; the conclusion was contained in the premises. The process by which such premises were formed lay outside the province of logic (Whately 1829, 230–1).[4] On the other hand, Bacon's methodology, although concerned with the formation of theories from factual data, regarded inductive generalizations as summaries of a class of observations. Whewell argued that neither of these approaches captured the mental process by which the major scientific theories were achieved. Writing later to Augustus De Morgan he was still making the same point, saying that if his interest in '*Discoverers*'

---

[4] Whately's *Elements of Logic* of 1826 rapidly became a standard text. I have referred to the treatise he wrote for the *Encyclopaedia metropolitana*.

Induction' did not have a place in mainstream logic, then '*so much the worse for Logic*' (Whewell to De Morgan, 18 January 1859, in Todhunter 1876, II, 417; Yeo 1985, 276).

In Whewell's account of science, the mind was dynamic and creative; great discoverers were imaginative and speculative in their quest for knowledge of nature. The true scientific attitude was not a passive contemplation of nature, as recommended by Goethe, but closer to Schiller's recognition of the 'authority of ideas over sense impressions' (Whewell 1840a, I, 34). Whewell suggested that the disagreement between the two German poets epitomized a long-standing philosophical choice, one in which psychological disposition played a significant part. To some extent he offered his notion of the 'fundamental antithesis' as a resolution of this. Both ideas and experience were necessary for knowledge of the external world. The crucial point was that the interdependence of these two elements – the ideal and the empirical, subjective and objective – began at the level of basic perception. Facts themselves involved 'an act of mind'; so did the inductive leap from disconnected facts to unifying theory. Thus 'fact and theory pass into each other by insensible degrees' (1840a, 45; see Buchdahl 1991 for a full discussion).

The first review of the *Philosophy* appeared in the *Athenaeum* (by De Morgan) on 12 September 1840. Ten days later Whewell replied in a privately printed pamphlet. Asserting the novelty of his approach, Whewell explained that Immanuel Kant's philosophy was preliminary to his special purpose. Kant showed the universal character of space and time as '*conditions of perception*'; but Whewell was concerned with the '*foundations of science*', and with the specific fundamental ideas that made '*universal and necessary truth possible*' in different branches of science (Whewell 1840b, 4–5). Yet the reviewer, he complained, had given insufficient attention to his account of other ideas, such as 'Force', 'Substance', 'Chemical Affinity', 'Symmetry', or 'Resemblance', that governed the perception of phenomena in particular sciences (see the list in Whewell 1847a, I, 80). Furthermore, in his paper on the 'Fundamental antithesis of philosophy', which led to a revision of the *Philosophy* in 1847, Whewell underlined the high degree of intellectual specialization required for the clear grasp of fundamental ideas in each branch of science (Whewell 1844). The implications of this will be considered in chapter 9.

Both Kant and Whewell were concerned with the nature of scientific knowledge. As Whewell expressed it, this involved the 'paradox of universal propositions'. How could laws, recognized as 'universal and necessary', be collected from experience by observation and experiment, as recorded in the history of science? Both thinkers sought to provide an epistemological justification for this knowledge (Whewell 1840a, 1, 238–91). The similarity of their answers was, as just noticed, a matter of debate as soon as Whewell's work appeared and is still being discussed. However, at another level there seems to be a distinction worth mentioning. In a more specific sense than Kant, Whewell was trying to legitimate science as an intellectual and social activity in a particular context – one he described as 'the emergency which has arrived' (1840a, 8).

In his famous essay, 'What is Enlightenment?', of 1784, Kant wrote: 'I have placed the main point of enlightenment – the escape from their self-incurred tutelage – chiefly in matters of religion because our rulers have no interest in playing the guardian with respect to the arts and sciences' (Kant 1949, 291). This notion of science as safe from political interference also figured at this time in the *Encyclopaedia Britannica* in a discussion about the gradual replacement of authority by experiment as a test of truth. This would continue to occur because the truths discovered by science 'are such as do not affect the passions of men'; those keen to keep people 'in political or religious ignorance' did not suspect this form of knowledge and, in any case, found it difficult to justify its obstruction (Robison 1797, 659). Whewell made a similar contrast, suggesting that while politics and morals immediately incited curiosity, the material sciences attracted interest of a 'less animated kind' (Whewell 1840a, 1, 15). This, he proposed, made them a good starting-point for an inquiry into the philosophy of knowledge – one that could eventually embrace the moral and social sciences.

Appeals to the neutrality of science have been part of the strategy followed by the Royal Society since the seventeenth century. At a time when political conflicts often turned on religious ones, the prospect of a body of knowledge governed solely by interrogation of nature was effective. While this still had its uses in the early nineteenth century, Whewell knew that, by itself, such a tactic was no longer an adequate means of promoting his preferred image of science. One reason for this was that science – as an activity and as a form of knowledge – was contested by various groups with different political and social values. Thus when he spoke of an 'emergency' and the need for 'reform', he was declaring his intention of showing that the success of the physical sciences must be understood within a philosophy of knowledge that corrected the errors of materialist and utilitarian doctrines. His alternative rejected the view – present in Locke's epistemology – that sensations were the basic element of knowledge, arguing that even simple perceptions required inference by an active mind. The theory of fundamental ideas – not innate ideas – emphasized the constructive role of the mind in science and the historical growth of knowledge. This philosophy of science, in Whewell's opinion, had implications for proper views of human nature and moral capacities.

INTRODUCING WHEWELL

Now that I have outlined Whewell's programme, and hinted at some of the aims behind it, it is important to offer some account of the position – social and personal – from which he launched this ambitious project. This immediately highlights the now well-recognized gap between the influential position he achieved and the marginal social background from which he came.

One point generally agreed about Whewell is that his rise from Lancashire to Cambridge was spectacular. Born on 24 May 1794, the son of a master carpenter, he came to Cambridge in the autumn of 1812 on a fifty-pound scholarship. His undergraduate career was sprinkled with rewards and prizes, although he had to be content with the position of second wrangler at his graduation in January 1816. His father died in July 1816, soon after this crucial moment in his son's life, but before he was elected to a Fellowship in the College on 1 October 1817. From here the achievements continued. In 1818 he became assistant tutor. In 1819 he published his first book, *An elementary treatise on mechanics*; in the next year he was elected as a Fellow of the Royal Society of London and in 1827 became a Fellow of the Geological Society. In 1825 Whewell was ordained as a priest in the Anglican Church – a necessary requirement for a college academic – and in the same year he was nominated for the Chair of Mineralogy, a position he assumed in 1828 after some disputes about the process of election were settled (Todhunter 1876, 1, 32; Stair-Douglas 1881, 101). He resigned this Chair in 1832 and, with very little sense of dissonance, took up the Chair of Moral Philosophy in 1838. Shortly after marrying Cordelia Marshall in 1841, Whewell was appointed to the Mastership of Trinity College in December of that year. The young man who wrote to his father in 1814 reporting that he had purchased a copy of Newton's *Principia* – 'a book that I should unavoidably have to get sooner or later' – now had full access to some of Newton's own books in Trinity College Library, and resided in the Lodge once occupied by Richard Bentley (Whewell to his father, 18 January 1814, in Stair-Douglas 1881, 11).

This is regarded as one of the great success stories of the period, but there is a danger in concentrating on the result – the picture of Whewell as the indomitable magus of Victorian science – without appreciating the tensions and insecurities associated with this social passage (Roach 1959; Rothblatt 1981, 31–4). Jonathan Hodge has recently reminded us of the social route which Whewell travelled: from Lancaster to Cambridge, but not via Manchester (Hodge 1991). In other words, Whewell managed to move from the family of an artisan to a Crown appointment on the advice of a Tory Prime Minister without having to negotiate the clash between the values of the established, landed culture – in which Trinity College was a major stake holder – and the new industrial capitalism. To some extent this is due to the fact that he married into a family, the Marshalls of Leeds, who had already accomplished this move. John Marshall was a successful flax spinner whose children married into the aristocracy. The eldest daughter, Mary, married Lord Monteagle in April 1841. In the autumn of that year, Cordelia married Whewell (Rimmer 1960). It is a testimony to Whewell's social transformation that their marriage was seen as another of these suitable alliances.

Whewell's perception of himself as a person extending the limits normally associated with his social background was already acute in his early student days. His letters from Cambridge to his family convey a strong sense of being their chosen representative in a different social world. They tell a story of a young man who regarded himself as the bearer of the high aspirations of the family in Lancaster. Announcing a first placing in every subject in June 1814, Whewell told his father that '*We* have reason to be proud.'[5] In December of the following year he warned against undue expectations, asking his father not to assume that 'I am very confident of all the success I could wish – because nothing can be further from the real case, and I have great reasons not to be so' (December 1815). Whewell's aim was not

---

[5] Whewell to his father, 26 June 1814, WP, Add. MS. a. 301²; my italics. Copies of some of these letters have only recently been donated to Trinity College. The following references in this chapter to this set of manuscripts are given simply by the date of the letter.

just academic success, but a Fellowship – 'the most important object' – which would justify the financial support of his father and give him an independent position. His father died before the Fellowship was conferred, but Whewell explained to his sisters that it was 'the most substantial benefit at which you ever had to rejoice with me. It secures me a comfortable establishment for life at least so long as my life is a simple one' (1 October 1817).

The social significance of this achievement was central to Whewell's understanding of his efforts. When his father became ill in May 1816, he told his sisters that their father was a model to be emulated: he hoped that they could all fill their 'situations in society, whatever it may be, as respectably as he has done'. But Whewell's goals went beyond this: 'And I yet hope that he will see some of us fixed in situations which he never could have aspired to, if it had not been for his indulgence and his efforts' (21 May 1816). Whewell's success, in his own estimate, lay in the fact that he could imagine higher goals than the father who was not 'allowed to see what he has done for his son' (14 August 1814).

It is not surprising therefore, that Whewell had made contact with the landed classes well before his marriage. In 1823 he met Sir John Malcolm who lived at Hyde Hall near Cambridge and subsequently enjoyed the friendship of the Malcolm family, to whom he introduced Hare and Sedgwick. There is a definite sense in which this was a surrogate family for Whewell (see Clark and Hughes 1890, 1, 282–3). His mother died in 1807 (when he was thirteen), three younger brothers by 1812, his father in 1816, and his sister Elizabeth in 1821. From this time, Whewell's contacts with his own family were limited to correspondence with two sisters, Martha and Ann. Many of the remaining letters explain his inability to visit Lancaster (27 September 1819; 16 January 1820). Moreover, there is a clear contrast between the letters to his father and sisters as a young undergraduate, when he was pleased to have them share in his academic triumphs, and the unease at the time of his wedding to Cordelia Marshall, a ceremony which none of his family attended. His letters to his sisters in the years after the wedding list his increasing integration with the aristocracy, almost rationalizing his belief that his family would have felt uncomfortable at the wedding in 1841 (Brooke 1987, 22; Rimmer 1960, 221).

Whewell's progress to the centre of English social life was not accomplished without other personal struggles. One example, which appears in some of the obituaries, was his social demeanour – one that transgressed the conventions of Cambridge life. An early manifestation of this was his English pronunciation and traces of north-country dialect. *The Athenaeum* related the story of his comment upon a herd of swine being driven past the College gate soon after his arrival: 'They're a hard thing to drive – very – when there's many of them – is a pig' (anon., 1866, 333). The accuracy of this account is less significant than the fact that it was recorded. Its interest is also reinforced by Whewell's own remark about a meeting with the French savant, J. B. Biot, in November 1817. Whewell was struck by the fact that his 'bad English and false accent gave you a superiority over him' (Todhunter 1876, 1, 353).[6] This was no doubt the observation of one who had suffered in this respect, but also that of a person who soon easily mastered the required usage. After all, this was the man who could later, as Master, have the following exchange with his Queen about the sewerage system of Cambridge. Queen Victoria (standing on a bridge over the Cam): 'What are all those pieces of paper floating down the river?' Whewell: 'Those, m'am, are notices that bathing is forbidden' (Raverat 1967, 34).

Another feature not so easily transformed was Whewell's personal style. He did not fit the available stereotypes sketched by Rothblatt in his study of the changes in academic life during the nineteenth century: he was not the genial but idle and indifferent don; nor was he the 'crabbed and ageing' pedant (Rothblatt 1981, 190–1). He did not anticipate the later, more activist role adopted by men such as Leslie Stephen and Henry Sidgwick, who revitalized the teaching and pastoral role of the university academic. The anecdotes about Whewell's threatening physical presence were a long way from the image of either the eccentric don or the later proponents of muscular Christianity (for the 'muscular' side of his behaviour, see Rothblatt 1981, 188). Among his friends, Whewell enjoyed a reputation for great charm and personal friendship, although he could also be blunt and domineering. In 1835 Joseph Romilly, the Registrar of the university and a Fellow of Trinity, recording a dinner with Italian guests, including 'Mr de Cavour', noted that 'Whewell was the life of the party.' But the same observer had also been struck some years earlier by another side of Whewell. In his diary for 7 May 1831 Romilly confessed: 'Committed unwittingly a cruelty on a Quadru-

---

[6] For the significance of standard English pronunciation, see Burke and Porter 1987, 2.

ped: I had engaged to go over in a fly with Whewell to Madingly:–
upon getting in at his rooms I found myself, Prof. Smith and Philip
Frere, and we were all 4 drawn by one horse!' (in Bury 1967, 77, 147).
This was not an isolated incident; in fact Whewell's horse-riding style
is a recurring theme in comments of acquaintances and close friends,
who regarded it as excessive (Schaffer 1991, 203). Airy remembered a
section of Cornwall as the place 'where Whewell overturned me in a
gig' (Airy 1896, 84).

The question of personal disposition was prominent in Whewell's
thoughts in the year preceding his appointment as Master. He wrote
to Hare in December 1840 asking for 'counsel' concerning the future
direction of his career. Feeling unease at the prospect of continuing at
Trinity with all his friends gone, he mentioned the possibility of
taking a College living. The idea of assuming the post of Vice-Master
did not suit him: as he put it, this carried 'authority enough to compel
you often to interfere, but not enough to enable you to carry your
point' (Whewell to Hare, 13 December 1840, in Stair-Douglas 1881,
207–8). Given his disagreements with colleagues about 'College
government', it was absolute power or nothing. In replying, Hare
confronted Whewell with recent reports of his unacceptable behaviour, saying that he had hoped that 'the vehemence of your nature
had been partly subdued'. But while Hare admitted that this might
be a reason for a move from College, he hoped that Whewell would
find a post more in tune with his abilities than that of a country cleric.
Mentioning a deanery or the Mastership, he suggested that
Whewell's proper ministry was that of a 'doctor' rather than a pastor
of the Church (Hare to Whewell, 17 December 1840 in Stair-Douglas
1881, 209–13). This advice allowed Whewell to see his talent for
'systematiz[ing] portions of knowledge' as a moral mission (see
chapters 3 and 7 below).

On 12 October 1841, the day of Whewell's marriage, Christopher
Wordsworth wrote to say that he was resigning and that he hoped
Whewell would be the new Master. Herschel and Hare both wrote to
influential figures on his behalf but Robert Peel, the Tory Prime
Minister, had already recommended his appointment to the Queen.
In a letter of 17 October he offered the position to Whewell, saying
that he would always be gratified to remember that he did this before
receiving any advice (Peel to Whewell, 17 October 1841, WP, Add.
MS. a. 78 no. 1; Stair-Douglas 1881, 225–7; Winstanley 1940, 81–2).

As Master, Whewell's style attracted more criticism, even from friends. F. D. Maurice described him to Hare in 1848 as 'arbitrary, unconciliatory and sometimes excessively rude' (Rothblatt 1981, 212). Apart from the objective facts about his behaviour, there may have been social reasons for the unease he provoked among students, and some Fellows. There is a clue in the *Athenaeum* obituary: 'His manners', it recalled, 'had not the polish which is only acquired in early youth' (anon., 1866, 334). There is an interesting illustration of the social prejudices Whewell aroused in an anonymous, printed letter addressed to him as Master from an 'undergraduate' in 1843. This charged that he was:

a rude, coarse, overbearing man, elevated out of all sense of propriety by a dignity, which to his birth and youthful fortunes appeared inaccessible; an unnecessarily strict and vexatious disciplinarian; ... intoxicated with elevation, he appears insanely to avenge the lowness of his birth upon all those who (by no fault of theirs) have been born above him. (Anon., 1843, 4)

The author went on to say that Whewell's achievements as a scholar and supporter of inductive philosophy were overshadowed by his reputation as an imperious bully. Rather than being an example of the productive way in which 'one of nature's aristocrats' could mix with 'our conventional aristocracy' – as Benjamin Franklin did – Whewell had made it less likely that other 'low born talents' would repeat his social rise (anon. 1843, 5–6).[7] In the light of this, it is significant that Whewell later opposed the idea of a separate College for poorer students, arguing that the advantage of university education was that 'by mixing persons of different conditions and destinations it gives them a fellow feeling and a mutual understanding as cultivated Englishmen' (Whewell to Edward Coleridge, 27 August 1843, in Stair-Douglas 1881, 297–9). But the evidence of Whewell's own experience suggests the difficulties involved. Although he made firm and lasting friendships with men from the upper and landed classes, the combination of a dominant personality and a lower-class background made Whewell's Cambridge career uncomfortable at times, both for him and for those who could not accommodate his surprising success. In spite of his assimilation, sensitivity about status is apparent in Whewell's response to Roderick Murchison's efforts to

---

[7] In 1838 a young Trinity man, John Lang, was expelled after circulating a blasphemous parody on the Litany including the petition that 'Mr Whewell may learn the manners of a Gentleman' (Bury 1967, 141–2). In his own student days Whewell was prepared to defy the authority of the Vice-Chancellor when, as President of the Union in 1817, he told the Proctors to withdraw before considering their directive to disband (Whewell to Rose 25 March 1817, WP, O 15. 47^{370}; Stair-Douglas 1881, 41).

entice him into the Presidency of the British Association, partly on the grounds that his Lancashire origins would appeal to the provincial audiences: 'I am *not* a man of science, ... *not* a man of popular manners' (Whewell to Murchison, 18 September 1840, in Todhunter 1876, II, 288).

Without a detailed biographical and psychological study it is hazardous to do more than notice that Whewell was, at least in his early career, a marginal person seeking to show that science was not a marginal activity. Other individuals, most famously Faraday, had come from poor backgrounds to the centre of the scientific world; but Whewell's case was different because, in making critical commentary on science his prime occupation, he had simultaneously to legitimate the metascientific role he assumed. In one sense this activity was associated with science itself, and other major figures made significant contributions. However, these men – Herschel and David Brewster in particular – were major scientific discoverers. Whewell did not see himself in these terms, and consequently, had to confront the idea of a metascientific role as *fundamental* to his vocation.

WHEWELL'S FRIENDS AND CRITICS

Whewell was not the only person aiming to define science. There were other major personalities, also significant participants in the debates on this issue, who regarded such commentary as part of their business. It is important to make acquaintance with these people now because they were both an elite audience for Whewell's philosophizing, and to some extent critics and judges of his performance. The men who most clearly fit this category are John Herschel (1792–1871), David Brewster (1781–1868), Baden Powell (1796–1860), Richard Jones (1790–1855), and Augustus De Morgan (1806–71). Their relationship with Whewell and his activities took different forms; but through their own books, reviews of his works, or personal correspondence, they constituted his key point of intellectual reference as well as sources of both inspiration and incitement.

John Herschel, the leading astronomer of the day, was one of Whewell's lifelong friends from his student days at Cambridge. He graduated as senior wrangler in 1813, when Whewell was in his second undergraduate year. Their correspondence extended from 1817, when Herschel left the university, to the time of Whewell's death, and covered a range of scientific topics, including the historical and philosophical project that Whewell embarked on in the 1830s. As we shall see in chapter 4, Herschel's *Discourse* of 1830 was an influential statement on the method and value of science; in some ways it was an example of the kind of inquiry Whewell wanted to pursue, while also representing views he subsequently rejected. Whewell's review of this was his first attempt to discuss the nature of science in a public forum. Herschel's extensive double review of the *History* and *Philosophy* in 1841 was not only an evaluation of these books, but an assessment of Whewell's vocation.

By contrast, Brewster was a declared enemy, at least for a period at the height of Whewell's career. Inventor of the kaleidoscope, and a major contributor to experimental optics, he stood for a view of science that was in many ways antithetical to Whewell's. As a Scot, Brewster was antagonistic to the increasing influence of English educational programmes in Scottish universities and, in particular, to the socially limited role for science allowed by Cambridge mathematicians such as Whewell, Peacock, and Airy (Davie 1964). As a practising scientist without a university appointment, Brewster gained an income from prolific journalism in the major periodicals, and by editing the *Edinburgh Encyclopaedia* from 1809 to 1830. Inevitably, given the scale of this activity, he reviewed the Bridgewater treatises, the *History* and *Philosophy* – all negatively – and, more notoriously, published a book in reply to Whewell's anonymous *Of the plurality of worlds* of 1853. This set off a public debate almost equal in intensity to that stimulated by the *Vestiges of the natural history of creation* (1844), published anonymously by Robert Chambers, and the *Origin of species* (1859). Whewell wrote letters to editors in response to some of these reviews and engaged Brewster over the plurality-of-worlds issue (Brooke 1977). Their confrontations were the most explicit in the metascientific debates of the period – natural theology, science policy, the role of technology in the history of science, the value and validity of a systematic philosophy of science – were all points of contention. Interestingly, they agreed on some things: the weaknesses of Bacon, the strengths of Newton and, significantly, on the necessity of commentary on science and the promotion of it in both expert and lay forums. Brewster was a practising scientist who entered metascientific disputes; Whewell was a scientific critic without practising as much science.

Baden Powell, like Whewell, was a mathematically trained university don and Savillian Professor of Geometry at Oxford, who took it upon himself to monitor scientific thought. Powell did not regard himself as an original thinker, telling William Rowan Hamilton in 1833 that he did not hope to contribute 'to the advance of Science in any but the very humble department of endeavouring to elucidate its first principles' (Graves 1882–5, II, 39). Pietro Corsi has recently shown how he did this with special reference to the impact on Anglican teaching, neglecting technical details in favour of general principles (Corsi 1988, 9–20, 276, 286). Like Brewster and Herschel, Powell commented on Whewell's works and saw them as a powerful illustration of the synthetic view of science that the age required, but he disagreed with the direction of Whewell's thinking in a number of areas.

Richard Jones was Whewell's closest adviser on matters concerning metascientific criticism. Their correspondence dates from 1817 when, as Whewell put it, Jones and Herschel 'vanished' from Cambridge leaving him with a sense of 'intellectual loss' (Whewell to Herschel, 6 March 1817, HP, Royal Society, vol. 18, no. 158). When the idealist direction of Whewell's epistemology became apparent, Jones did not endorse it; but by this time he shared the aim of understanding and disseminating the 'inductive philosophy' (Whewell to Jones, 23 September 1822, WP Add. MS. c. 52 no. 15). In turn, Whewell was Jones's mentor in his painful attempt to publish a book of non-Ricardian political economy. Discussions about this subject led to their earliest involvement with questions of scientific method and its moral and political connotations. In chapters 4 and 7 we shall see that Whewell's private communications with Jones allow an intimate view of the personal investment Whewell had in the philosophy of science.

Augustus De Morgan, one of Whewell's former students, was fourth wrangler in 1827 and Professor of Mathematics at University College London. He had a scholarly interest in the history of mathematics and science, and wrote on the Newton–Leibniz quarrel, on Newton's life, and on a wide range of scientific topics in the periodical press, especially in *The Athenaeum*, where he reviewed Whewell's major works. He also wrote more popular articles for the *Penny encyclopaedia*. Some of this journalism was collected in his eccentric and witty *A budget of paradoxes* – arguably one of the best examples of scientific satire.

The immense collection of letters in the Whewell Papers at Trinity College indicate the importance of private communication to his work. Leslie Stephen observed that Whewell maintained close friendships, and this shows in the candid nature of his correspondence. Whewell regarded his major works as contributions to a public discourse on the nature of knowledge, but much of his preliminary thought on these issues depended on continuing private discussion with friends and colleagues. So, too, did the sense of his public vocation as a metascientific critic. On this last point, his friendship with Jones – partly recoverable through the 572 extant letters between them – was crucial. Also important, although less extensive, was his correspondence with Hugh J. Rose, Julius Hare, William Rowan Hamilton, and William Wordsworth. On more specific issues Charles Lyell (1797–1873), Charles Babbage (1792–1871), and James David Forbes (1809–68) were important. Finally, Whewell's friendship with the geologist Adam Sedgwick (1785–1873), to whom he dedicated the *Philosophy*, was possibly more complex than we can know: few letters exist because of their joint residence at Trinity, but those which do suggest a tense relationship, at least after Whewell became Master.

A NOTE ON TERMS

Whewell's own capacity as a coiner of scientific terms makes this note obligatory. His introduction of words like 'scientist' and 'physicist' indicate the shifting 'geography of the intellectual world' in which he lived. While Whewell was preparing his major works a variety of terms – natural philosophy, physics, natural history, natural knowledge, physical science – were all used to denote the activity he studied. In the late eighteenth century, the *Encyclopaedia Britannica* suggested that 'natural philosophy' and 'physics' no longer comprehended 'the whole study of nature', and that 'natural knowledge' was now the generic term (Robison 1797, 637–42). One reviewer of the *Philosophy* reflected on the changing meaning of 'science' over the last 200 years – from demonstrative knowledge to knowledge of nature, distinguished from 'literature' and 'art' ([Ulrici] 1847, 3). The suggestion that 'science' was coterminous with natural knowledge brought hostile responses from critics of the British Association for the Advancement of Science in 1831, but this shift was well under way, even though Whewell's word 'scientist' did not replace 'man of science' in most usage until much later (Ross 1962). In 1867 W. G. Ward said that he used 'science' in the 'sense which Englishmen so commonly give it; as expressing physical or experimental science, to the exclusion of theological and metaphysical' (cited in Hankins 1980, 388). There is little doubt that Whewell's works, in advertising and defining the inductive physical sciences, contributed to this contraction of meaning. His professed aim, however, was to embrace moral and social studies as sciences.

The next chapter hardly mentions Whewell. This is in keeping with the two points of focus mentioned earlier: the wider debates on the nature of science in the early Victorian period and Whewell's involvement as a major actor. However, in spite of his formidable intellectual reputation, Whewell was a participant in a debate he did not always control. Along with other men of science, Whewell had to engage in discussion and promotion of science in a 'public sphere' inhabited by various speakers and different audiences (Habermas 1974). Chapter 2 attempts to sketch the characteristics of this forum for public debate. An appreciation of the place of science in this makes it possible to understand why men of science in this period were concerned not only with substantive scientific theories and discoveries, but with a range of metascientific issues. Some of these – issues of method, epistemology, history – have since become the objects of specialist disciplines; in the early Victorian period they were matters of wider interest. This is not incompatible with the fact that some individuals paid particular attention to such issues. Thus chapter 3 considers Whewell as a person whose sense of vocation was informed by the urgency of this metascientific task. I do not suggest that he was the only individual pursuing such questions – he had critics and competitors, as mentioned above. The adoption of this role did not mean that there was a pre-defined *genre* in which Whewell could present such metascientific reflection. The next four chapters indicate some of the conditions and constraints on Whewell as he wrote and spoke on science in various levels of the public sphere in early Victorian Britain.

In Part Two, the four chapters consider some of the major issues canvassed by Whewell in his efforts to define science. These follow a chronological order, and to some extent are centred around his major writings, from the early periodical reviews (1831–4), the Bridgewater treatise on *Astronomy and general physics* (1834a), the *History* (1837), and the *Philosophy* (1840). Each of these chapters also reflects on the different form or genre of these writings, and on the relationship of his ideas to that of other commentators. Thus chapter 4 examines the crucial medium of the Victorian periodicals, the place of science in them, and Whewell's use of the review essay in his earliest attempts to raise metascientific issues before the educated public. Chapter 5 considers the way in which his work posed questions about the moral and intellectual character of the man of science; although he did not write a separate biography, Whewell's views connected with contemporary debates about the role of great men of science and the value of biographies as a way of understanding and presenting science. Chapter 6 looks at Whewell's *History* as a major vehicle for his views on the method of science, the role of great discoverers, and the nature of scientific progress. It also considers the ways in which he used or modified the concepts of existing scientific and political historiography – such as progress, revolution, and tradition – to present a new, non-Baconian account of the history of science. Reviewing some of the previous discussions, chapter 7 analyses the long-term concerns that led Whewell to write the *Philosophy*, arguing that although this is now seen as one of the foundational texts of the philosophy of science, its genesis lay partly in concerns apparent in his earlier thinking and writing on morals and political economy. Whewell's interest in a wider philosophy of knowledge embracing both physical and moral science continued after 1840, but from this time epistemological issues began to become separated from the wider context in which Whewell conceived them. Paradoxically, his own *Philosophy* may have hastened this divorce, even though Whewell advocated his idealist epistemology as an element of moral reform, and as a justification of his metascientific vocation.

In Part Three, the final two chapters consider the application of Whewell's account of science in two areas. This requires some return to the earlier part of his career, but both chapters also take the discussion up to the 1860s. Chapter 8 deals with his educational writings, especially the publications of 1837 and 1845 on university education. Here Whewell sought to extend his theories of scientific knowledge more generally to the nature of liberal education, traditionally dominated by mathematics and classics, and also addressed the question of the place of science in the university curriculum. When taken with his other writings of this period on the place of science (and technology) in society, we can see how his views on science had implications that did not always please the scientific community. Chapter 9 discusses another area of application: the implications of Whewell's philosophy of science for studies of social and moral science and its bearing on natural theology and values. The interest here is in the way in which Whewell's views about the nature of scientific knowledge encouraged extrapolations to the social and moral sphere. In resisting some of these, Whewell was out of step with the drive towards synthetic systems encouraged by Auguste Comte, the other major metascientist of the period. Whewell's philosophy of science created possibilities for natural theology while at the same time complicating the approaches of some other commentators on science. The pervading issue here is the meaning of the term 'unity of science'.

CHAPTER 2

Science and the public sphere

In the late nineteenth century, the intellectual historian John T. Merz wrote that: 'It will be generally admitted that the scientific spirit is a prominent feature of the thought of our century as compared with other ages. Some may indeed be inclined to look upon science as the main characteristic of this age.' In saying this and then devoting a major part of his work to the nature and development of natural science Merz was confirming its emphatic presence in the culture of that time. But he went on to state that it was 'not necessary to *define* what I mean by science', thus recording the distance between his position in the 1890s and that of writers earlier in the century who regarded the definition of the aims, methods, and cultural status of natural knowledge as a crucial and difficult task (Merz 1896–1904, 1, 89).

My aim in this chapter is to set out a framework in which these early Victorian debates can be understood, to discuss the intellectual and social conditions that sustained a discussion not only of scientific discoveries but of the nature of the scientific enterprise. In this period there was a wide-ranging debate on the nature of science covering topics which are now the focus of separate, professional areas of study. Thus John Herschel could discuss quite detailed points of scientific method in his *Preliminary discourse on the study of natural philosophy* (1830), a book published in a popular *Cabinet cyclopaedia*. While it has been recognized that this was an apologia for science, it is often Herschel's allusions to the harmony of science and religion that have been noticed, rather than the way his book made questions of method and philosophy central to a definition of the scientific vocation conducted in public as well as private forums. While Herschel's text has been successfully mined by philosophers of science seeking to recover the origins of their discipline (Laudan 1968), we also need to regard these topics as crucial to the contemporary legitimation of science (Agassi 1971; Schweber 1981; Yeo 1981 and 1989).

Susan F. Cannon, one of the most important scholars of the cultural history of Victorian science, sometimes gave the impression that the meaning of science was clear, and not in need of further definition: to be 'scientific', as she said, was to be like John Herschel (Cannon 1961, 238). This was because she was largely concerned with showing how science was justified by its connection with religion, through natural theology. But the undoubted importance of this rationale does not imply that science required no further definition or defence. Thus when Thomas Young agreed with Herschel in lamenting that 'much is *nicknamed* science', he was referring to other concerns about the cultural meaning and status of natural knowledge (Young to Herschel [1828] in Hall 1984, 41). Cannon gave little detailed attention to the methodological, epistemological, historical, and biographical issues that comprised this debate. Writing on the late-nineteenth-century conflict between the new scientific naturalism and traditional 'clerical science', F. M. Turner suggested that this involved an 'epistemological controversy' (Turner 1978, 359). But we should not assume that such issues were absent in the earlier period when natural theology was not such a direct target. The range and subtlety of this discourse has often been overlooked, partly because the existing historiography has underlined the way in which science was integrated with early Victorian culture through natural theology.

SCIENCE AND NATURAL THEOLOGY

Since the influential work of Cannon, dating from the early 1960s, there has been a consensus that it is profitable to speak of 'science in culture' rather than about 'science and culture'. In making this point Cannon was arguing against the blinkered perspective produced by studies that approached the early nineteenth century with the categories of twentieth-century disciplines. This had led to a situation in which science was treated separately from social and political issues, and literary debates were allowed to stand as the unchallenged representative of cultural activity (Cannon 1978, 257). Cannon's response to this assumption involved the strong claim that in the early Victorian period science functioned as a 'norm of truth'. In a stimulating essay of 1964 she tabled examples from writers such as Chalmers, Wordsworth, Newman, and Ruskin to illustrate their obeisance to the conclusions of science as the standard of truth in human inquiry. The impression given here was that everyone knew what science was and that most agreed on its high intellectual status. Cannon made special use of the fact that even John Henry Newman, not a supporter of science or scientists, had cited Newtonian laws when he wanted an incontestable example of truth. But in her own examples it is possible to see signs of tension in attitudes to science: Newman referred to astronomy and not to other sciences, Wordsworth contrasted mineralogy and geology and substantially qualified Humphry Davy's paean to the moral and social merits of science, and Newman also rejected the arguments of natural theology that gave scientific inquiry a religious foundation (Cannon 1978, 6–9, 13, 20).

Cannon argued that such tensions became significant after 1860; until then they were constrained by what she called the 'Truth Complex' – a set of assumptions grounded largely in the discourse of natural theology as it developed from the seventeenth century via the moderate English Enlightenment. Its organizing component was the premise that all truths – theological, poetic, and scientific – were in harmony, that there was essentially only one truth. This unity was shattered by Darwin's *Origin of species* (1859) when it severed the link between what Sedgwick called the 'material and moral' spheres by outlining a picture of nature that made inferences to a benevolent Deity implausible. But it was not only the theological attack on Darwin that undermined the 'Truth Complex': Cannon pointed to the way in which scientific criticism of the theory of evolution by natural selection released a number of conflicts within science itself, such as the clash between geological reasoning and the kind of evidence from physical astronomy about the age of the earth advanced by Lord Kelvin, and the statistical arguments put by Fleeming Jenkin. This meant that there was now no agreement on questions of method, evidence, proof, or truth, in science; hence it could no longer provide a stable foundation for the unifying vision of natural theology.

On this view, the close integration of science within a Christian culture existed, almost naturally, as a legacy of the eighteenth century; its moment of crisis did not arrive until the appearance of Darwin's theory. Although disputing Cannon's explanation, other writers have generally accepted the view that the test posed by science came in the second half of the century. For F. M. Turner this appeared as the young generation of scientific naturalists such as Huxley, Tyndall, and Spencer fought to displace the earlier 'clerical scientists' and their dependence on religious apologetics (Turner 1978). Robert M. Young differs in so far as he stresses the continuity between the older natural theology and the new scientific naturalism: both were attempting to find 'ways of rationalizing the same set of assumptions about the existing order'; that is, both sought natural justifications of the structure of capitalist society (Young 1985, 191). But Young shares the view that the cultural standing of science in the first half of the century was comfortably assured by its connection with natural theology.

I do not intend to question the importance of the intellectual and apologetic framework supplied by natural theology. But the emphasis on it as the chief rationale for science may have created an erroneous sense of the security of natural science in early Victorian culture. A second look at this issue is useful because it opens another area of inquiry: namely, the character, forms, participants, and audience of what I have called 'metascientific' discourse. This embraced more than the standard arguments from natural theology, and included discussions of the method and moral character of men of science, the history of scientific discovery, the hierarchy of its separate disciplines, the application of scientific concepts and reasoning to other areas, and the appropriate means of explaining science to different audiences.[1] The significance of this can be better appreciated when we recognize that science was a relatively insecure cultural activity.

There are at least two sources for this revision. One derives from studies of natural theology itself. John Brooke and Pietro Corsi have qualified some of the assumptions about the harmony between science and religion in the first half of the nineteenth century, epitomized for Cannon by the Broad Church and the Cambridge network. Corsi offers a more fine-grained analysis of the tensions and strategies beneath the surface of natural theology and its effort to accommodate scientific knowledge within the intellectual framework of Christian theology. Tracing Baden Powell's career from the 1820s to the eve of the *Origin,* he confirms what Brooke discerned in the plurality-of-worlds debate of the 1850s: namely, that natural theology was in danger of fragmenting before the impact of Darwinian thought, as rival religious denominations sponsored conflicting interpretations of scientific theories (Brooke 1977, Corsi 1988).

---

[1] Debates now seen as part of philosophy of science did indeed involve moral considerations related to natural theology; but this requires an extension of Cannon's perspective. See chs. 5, 7, 9 below.

Starting from the seventeenth-century background, John Gascoigne has shown that changing institutional forces within Cambridge towards the end of the eighteenth century made the previous 'holy alliance' between science and religion a less reliable part of the intellectual landscape (J. Gascoigne 1989). These studies suggest that there was no easy defence of science, even in the early nineteenth century. This is borne out by the anthropological scrutiny of the early years of the British Association for the Advancement of Science, provided by Morrell and Thackray, which shows that any consensus about the place of science in culture was one that had to be carefully constructed (Morrell and Thackray 1981).

The second source is the social and institutional history of science. Here there is a different approach to the changing status of science over the century. For Cannon and Young, the decade following Darwin's book marks the demise of a period in which science was a central component of cultural debates, such as those turning on the question of man's place in nature. But in sociological and institutional terms, the 1860s are identified as the *beginning* of the autonomy of science, marked by the greater availability of scientific careers and greater support from the State. In the work of Turner on scientific naturalism, and that of MacLeod and Alter on the relations between science and the State, we find a deliberate elaboration of a professional scientific identity, often in conjunction with an attack on the unscientific attitudes and classical education of the political establishment (Turner 1978; MacLeod 1972; Alter 1987, ch. 5). This aggressive confrontation contrasts dramatically with the style of the early British Association and its Cambridge and Oxford managers. It was certainly not the way in which William Buckland and Whewell courted Sir Robert Peel. In short, after about 1870, there was a scientific culture, rather than science in culture. The view from this vantage-point is useful because it allows the suggestion that science in the early part of the century was a marginal activity that had to be defined, thus producing a level of metascientific debate.

'SCIENCE' AND MEN OF SCIENCE

In the early 1800s, science did not enjoy the cultural and institutional security that later allowed Merz and others to see it as the dominant feature of the century. Its prestige was lower than rival forms of intellectual activity, such as theology and the classics, which, even if they did not attempt to explain the natural world, stood as powerful exemplars of culturally sanctioned bodies of knowledge. In spite of the popularity of some sciences in the early nineteenth century, there is evidence of a distance between scientific inquiry and other intellectual pursuits. For example, Caroline Fox recorded in her diary for 1840 that: '[John] Sterling is taking up geology as a counter-current for his mind to flow in, a subject so removed from humanity that he considers it one of the least interesting of human sciences' (Pym 1882, 11, 60). Thus the facility with which men of Cannon's Cambridge network moved between science, natural theology, political economy, and German literature does not necessarily mean that all these subjects had equivalent cultural status (see Rupke 1983 on geology).

The word 'science' had not entirely lost its earlier meaning of systematic knowledge, or *scientia* – for some people, logic, theology, and grammar were still 'sciences', and the term was still used synonymously with 'philosophy', even though the British Association attempted to make 'science' mean natural knowledge. 'Natural philosophy' could refer generically to the study of the natural world, or more specifically to those inquiries that manifested the quantitative and experimental approach exemplified by Newton; hence it could invoke the authority of the traditional concept of 'science'. On the other hand, 'Natural History' was just beginning to escape from the low status it occupied in the early eighteenth century. Leslie Stephen remarked that in this period it 'had been regarded with good-humoured contempt as a pursuit of bugs, beetles and mummies... Now it was beginning to be recognized that such pursuits might be a credible investment of human energy' (cited in J. Gascoigne 1989, 233).

Yet in one sense any specific definition of these two general terms – natural philosophy and natural history – was made redundant by the new vocabulary of specialized subjects. Between 1781 and 1840 the monopoly of the Royal Society was overthrown by the foundation of some two dozen specialist scientific societies (Morrell and Thackray 1981, 546; Alter 1987, 14–17; Emerson 1988 on the Scottish scene). By the early nineteenth century, the entries on these general terms in major encyclopaedias were displaced by detailed articles on separate disciplines; if included at all, they invariably pointed the reader to the entries on specific subjects (Yeo 1991b). Similarly, the people who studied the natural world were coming to be identified as astronomers, chemists, botanists, or geologists. Babbage complained in 1851 that 'our language itself contains no *single* term' by which the occupation of men of science could be expressed (Babbage 1851, 189). Whewell made this point in 1833. The phenomenon they observed meant that the definition of 'science' – as distinct from particular disciplines – was a serious problem.

This indefinite vocabulary does not in itself accurately indicate the status of science; but it does offer an alternative intepretation, within a chronology similar to the one employed by Cannon and Young. In contrast with its position in the late nineteenth century, science in the early Victorian period was relatively weak, in spite of its connection with natural theology. This does not imply its insignificance, but rather highlights a set of social and intellectual conditions that required and encouraged a broad consideration of the nature of scientific practice. It is worth commenting separately on these two factors – the social and the intellectual.

In his lecture on 'Science as a vocation' of 1919, Max Weber contrasted the clear scientific careers then available in Germany and North America with the older notion of an 'inner vocation for science' (Weber 1989, 8). But in the early nineteenth century only the latter was a real possibility. This is registered in the well-known cry from Babbage in 1830 that there was no such thing as a career in science, and the complaint of the Declinist campaign that this was especially true in Britain (Babbage 1830, 14–39). It is now well recognized that there were significant differences in the manner in which individuals came to be scientists. Without some independent means of support, either from inherited wealth or a university or clerical position, a regular and substantial commitment to *research* in science was almost impossible. Brewster advised the young James Forbes in 1830 that a well-managed legal career offered the best chance of sufficient money and time for scientific research (Shapin 1984, 18). Michael Faraday, the most famous experimental scientist of the period, was fortunate to hold one of the few specific research positions, as Director of the Laboratory at the Royal Institution after 1825. But contemporaries saw him as exceptional, partly because of his poor background. Writing to Babbage in 1835, Lord Somerset commented on Faraday's 'conjunction of poverty and passion for science', describing it as 'romantic' (Berman 1978, 174).[2] Indeed, it has been observed that,

---

[2] But see Cantor 1991b, 108, for Faraday's secure financial standing by this date. Cultivation of science as a member of a scientific society was relatively expensive: for example, three guineas per year for Geological Society membership. See Rudwick 1985, 481.

on the whole, both women and lower-class men were excluded from the social circumstances that allowed science as a personal vocation (Abir-Am and Outram 1987, 2–4). This meant that the choice of a life in science had to be justified by every 'man of science'. One result was that metascientific reflection was not separate from the scientific enterprise; it was an essential condition of its public recognition.

This social insecurity of the scientific enterprise coexisted with some major intellectual developments. The significance of these has led some writers to speak of a 'second scientific revolution' in the period between the French Revolution and the decade following the close of the Napoleonic wars. This has been variously conceived. Thomas Kuhn wrote of the 'mathematization of Baconian physical science' as an important alteration of the earlier intellectual land-scape divided between the Newtonian sciences of astronomy, mech-anics, and optics, and the array of observational subjects usually classified as parts of natural history (Kuhn 1977, 220). Some contemporaries believed they were witnessing revolutionary events in the intellectual sphere that paralleled those of the political (Coleridge 1817; Cohen 1985). We can now interpret these events as the collapse of natural philosophy and the formation of a new intellectual geography encompassing Lavoisier's chemistry, Cuvier's and Lamarck's biology and physiology, Laplacian physical astronomy, Fresnel's and Young's wave theory of light, the geology of Hutton and Lyell, and Faraday's electromagnetism. Others have seen the crucial shift at the organizational level: Roger Hahn underscored the growth in specialist scientific institutions and journals as constituting a 'second scientific revolution in the early nineteenth century' (Hahn 1971, 275; Cantor 1982; Cohen 1985, 92). The link between institutional and intellectual developments was made by the so-called 'Declinists', such as Babbage and Brewster, who pointed to the fact that many recent scientific advances occurred in France, where the financial support for savants was far more regular than in Britain.

This national comparison deserves comment, because it relates to the question of why science was discussed in the public sphere. It has been noticed that the appeals by men of science to natural theology were less obvious in France than in Britain during the early decades of the century (Crosland 1975, 8; Knight 1986, 30–1). Apart from the different religious traditions and political arrangements, one possible explanation of this contrast is that there was less need for public defences of science in France. Given the greater availability of

scientific careers in France, elite savants, such as those of the Paris
Académie, may have been able to withdraw from such apologetics.
But this interpretation is misleading if it is taken to imply that the
place of science in public debate in Britain was only necessary because
of the absence of professionalization. Inferences from the French
scene need to be made with care: the relative lack of natural theology
does not indicate that science was immune from public consideration.
Georges Cuvier, one of the most powerful figures in French science,
could not afford to ignore the public attitude to scientific ideas and
procedures. As Dorinda Outram argues, his conflicts with Jean
Baptiste Lamarck and Franz Josef Gall reveal a concern with the
proper conduct of science, the grounds of its authority, and its relation
to different audiences. His attempt to draw boundaries between good
science and bad popularization should not be interpreted as a sign of
professional consolidation; instead, this might reveal his appreciation
of the insecurity of the elite that sought to control the uses of science in
Napoleonic France (Outram 1984, ch. 6).

While it is true that some areas, such as the mathematically
dependent parts of astronomy, mechanics, and optics, were not
accessible to the layperson, the *idea* of scientific knowledge and its
intellectual claims, were part of a public debate, both in France and
Britain. In fact, there was awareness among contemporaries that
science might be affected by the different political and social
conditions of the two countries. This is illustrated by two articles in
the *Edinburgh Review*, both making connections between the advance
of science and public opinion about it. John Playfair pondered the
effects of the French Revolution on science, noting that State support
for scientific institutions was associated with increased emphasis on
utility. While this created a wider audience for science, it also meant
that the public needed to be educated in the history of science and its
lessons about the importance of free inquiry (Playfair 1810, 396–8).
Ten years later another reviewer circulated a series of stereotypes
favouring English over French science, and claiming that scientific
knowledge was more broadly diffused in England where, consequently, there was a less dramatic gulf between 'the learned and the
unlearned' (Chenevix 1820, 409; also 388–9, 415–17).

These examples indicate that there was an important public debate
on the nature of science in which notions of audience, communication, and public opinion were crucial. Furthermore, it was one closely
conducted within the scientific community, such that even those

urging a separation between science and socio-political issues were inevitably involved. Although, in Britain, general acceptance of the valuable rationale supplied by a broad natural theology continued to exist, the aims, method, and cultural implications of science were not subjects that could be left unstated, and when stated did not always produce agreement. With this, there was a strong sense in which the communication of scientific results was not divorced from advocacy of science itself. When Playfair acted as public-relations officer for James Hutton's *Theory of the Earth* – explaining it in his own *Illustrations of the Huttonian Theory of the Earth* in 1802 – he understood that novel approaches and theories in science would not simply speak for themselves. Thus when Lyell urged Darwin in 1838 to accept the necessity for philosophers to turn public orators, he was recognizing one of the conditions which governed scientific life in the period (Lyell to Darwin, 8 September 1838 in Lyell 1881, II, 45). Lyell's sensitivity to this was no doubt heightened by his effort to define a cultural space for geology, a subject without a solid mooring in the university curriculum (Rupke 1983; J. Gascoigne 1989). One aspect of this was his awareness of the need to make it accessible to a large audience – 'I must write what will be read', he told Gideon Mantell in 1826. He confided to Herschel that the successful sales of the *Principles* vindicated his attempt to write 'in a popular style' (Lyell to Herschel, 1 June 1836 in K. M. Lyell 1881, I, 464; Porter 1982, 48).

This issue of the appropriate communication of science overlapped with more general questions about authority, both intellectual and political, in the period before the First Reform Bill. Diagnosing the 'spirit of the age' in 1831, J. S. Mill spoke of an interregnum between the demise of traditional authorities and the emergence of a new intellectual class capable of teaching the public to discriminate between common sense and informed judgement. Several writers regarded the popularization of knowledge as a problem, and in 1833 Edward Bulwer-Lytton was quite specific in stressing the need for what he called the 'government of knowledge' (Bulwer-Lytton 1833, II, 122).[3] The problems in the case of science were serious. Writing in the *Quarterly Review* in 1829, Robert Southey listed a range of sciences, some of which had been recently 'created or exceedingly extended',

---

[3] Alexis de Tocqueville's *Democracy in America* of 1835 connected political and intellectual authority. For other treatments of the notion of 'public opinion' in this period, see Yeo 1984, 7–8. On the link between public understanding of science and democratic values, see Ezrahi 1990, 81.

and concluded that it was impossible for the lay public to master them. This comment was part of the debate on the 'education of the people' in which Brougham and some utilitarian writers advocated the extension of science to the working classes. Southey and other conservatives predicted that this would result in the spread of superficial knowledge and the loss of proper respect for profound thought (Brougham 1826, 197; Bowring 1827; Southey 1829, 496–7). This point was accepted by some men of science. The advancement of science, as Forbes warned members of the British Association, was not always guaranteed by its diffusion to popular audiences, especially if these were more attracted by the utility of scientific discoveries and by factual results rather than by theoretical implications (J. D. Forbes 1834, xii–xv). Science needed an audience, and its proponents had to make claims about its social relevance, but there was a danger that the image produced might not be one desired by the leaders of the scientific community (Yeo 1981, 75–8).

There is some doubt whether 'popularization' is an adequate term to describe the public discussion of science that occurred in this period. In the twentieth century, popularization is conceived as a communication of expert scientific knowledge to a lay audience; but this notion of a clear distinction between expert scientist and lay audience is inappropriate in the early Victorian period (Shapin 1990b, 991–3). Many accounts of science, in periodicals and encyclopaedias, were aimed at practising scientists as well as a wider, lay readership (Yeo 1991b). Moreover, in the first half of the nineteenth century public discourse on science served at least two purposes: it not only conveyed scientific discoveries to the public, but also legitimated science as a part of cultural discourse. Where and how was this articulation of science and its values accomplished?

VICTORIAN PERIODICALS AND THE PUBLIC SPHERE

From the early 1970s an influential answer was provided by Young's notion of a 'common intellectual context', supported by natural theology, capable of embracing discussion of science and its bearing on the pervading question of 'man's place in nature'. By locating this in the major Victorian periodicals, Young gave more specificity to Cannon's idea of a 'Truth Complex' and was able to indicate the institutional conditions of the intellectual dialogue between science and other cultural activities. Thus while agreeing with Cannon that this dialogue began to dissolve by the 1870s, Young put the impact of

Darwin in a wider framework, pointing to the way in which scientific thought was now discussed in new specialist journals, such as *Nature*, *Mind*, and *Brain*, and less so in the leading general periodicals that constituted the forum of the common context in the earlier decades of the century (R. M. Young 1985, 127–8). This leads to the option of an account that places less emphasis than Cannon did on the impact of Darwin, and more on factors such as the extension and specialization of knowledge that made a general debate less feasible. But Young endorses Cannon's belief that natural theology provided the common framework for this assimilation of science within general cultural discourse – that is, until it became 'one point of view in a conflict' (1985, 135).

Writing in 1894, Huxley believed that this conflict was obvious by 1869, when *Nature* was founded. Men of science, he said were living under a Baconian rubric of two spheres – the religious and the philosophical – and were 'called upon to be citizens of two states, in which mutually unintelligible languages were spoken' (Huxley 1894, 1). This comment might confirm the contrast with the earlier period in which the general quarterlies carried reviews of science to a readership far wider than the scientific community. But Huxley also remarked that by this time it was 'rare even for the most deservedly eminent of the workers in science to look much beyond the limits of the speciality to which they were devoted'. This suggests that the notion of science fitting seamlessly into a common intellectual dialogue requires some qualification.

Without doubting that early-nineteenth-century readers could pass from articles on religion or literature to those on science, it is important to recognize that science posed some difficulties for a common discourse. Such a debate relied on a common language into which the technical concepts of particular subjects could be translated. Since these major quarterlies attracted an educated readership, this was usually possible (for estimates of circulation, see Gross 1969, 2; R. M. Young 1985, 154). Even in the case of political economy, where a set of technical terms was unavoidable, public discussion was possible because it concerned the subject as a whole rather than the complex differences between various doctrines (Checkland 1949, 41). But there is evidence that science was seen as a threat to this assumption of a common language. Herschel acknowledged this in his *Discourse*, noting that 'Science, of course, like every thing else, has its own peculiar terms, its idiom of language.' While it was not possible to do without this, he feared the effect of anything that 'tends to clothe it in a strange and repulsive garb' (Herschel 1830a, 70).

One illustration of this concern occurred in an article entitled 'On the application of the terms Poetry, Science and Philosophy' in the *Monthly Repository* (anon. 1834). The title itself indicates a self-conscious awareness of different intellectual fields. One of the questions raised was that of the appropriate language for science: should there be an addition of new words to express the increasing number of scientific concepts, or a more exact discrimination in the meaning of terms already in use? The author was convinced that the first alternative should apply: 'In case of a new science, such as chemistry or mineralogy, the objects and relations of which lie altogether beyond the ordinary circle of thought and observation, the former of these expedients is resorted to' (anon. 1834, 323). That is, the concepts of the various sciences were seen as removed from ordinary experience and hence in need of a special language. This view coexisted with the conviction that the word 'science' could still be applied to non-physical knowledge. The article contrasted science and poetry, but saw no reason why there should not be a 'Science of metaphysics, of morals, of jurisprudence, or of political economy, as well as of astronomy, mechanics and chemistry' (anon. 1834, 329). These specific reflections on the difficulties of a common language, at a time when the 'common intellectual context' should have been strong, call for some additional examination.

The concept of the 'public sphere', advanced by the German political and social theorist Jürgen Habermas, may be useful here. This concept attempts to encapsulate the region between the State and civil society that began to emerge in France, Germany, and Britain during the eighteenth century. As Habermas explains it, this was a new sphere of debate in which the rising bourgeoisie could criticize the behaviour and values of an absolutist state. Unlike participation in the political and social domains, engagement in this activity was supposedly not dependent on birth or position or privilege, but rather on the acceptance of universal norms of reason (Habermas 1989; Hohendahl 1982). In fact, it was a conversation between groups with both property and power – the landed aristocracy and the bourgeoisie. Habermas sees the operation of this public sphere as first discernible in Britain, partly because of the collapse of absolute monarchy and the alliances between these two classes embodied in the Restoration settlement of 1688. Periodicals such as Steele's *Tatler* and Addison's *Spectator* functioned as consolidators of a cultural consensus between these groups, and supported a critical discourse about literature, science, politics, and philosophy. Addison said that the aim was to bring 'Philosophy out of the Closets and Libraries, Schools and Colleges to dwell in Clubs and Assemblies, at Tea-tables and Coffee Houses' (Wolf 1938, 41). The editorial style of these journals, and the publication of books by subscription, encouraged the notion of a community of authors and readers (Eagleton 1984, 29). This found tangible form not only in clubs and coffee houses, but in lending libraries and literary and philosophical societies that emerged in England during the eighteenth century. Following Habermas, Peter Hohendahl sees all these as part of a literary public sphere in which the reading public was the arbiter of taste and aesthetic standards. This represented a definite break with the former assumption that such judgements were the province of aristocratic, courtly society – they now belonged to a discourse among educated, private individuals, and this was the basis of a political, as well as a literary, public sphere (Hohendahl 1982, 52–3; Habermas 1989, 51–2).

How can this help our understanding of the context in which debates on science occurred? To begin with, it allows a historical perspective on the discourse conducted in the Victorian periodicals, one that situates them in relation to changes in the public sphere from the late eighteenth century. Habermas emphasizes that from its inception this realm of public discourse was based on the notion of an identification of 'property owners' and 'common human beings', that is, on the fiction of *one* public. This, he explains, had positive effects for the 'emancipation of civil society from mercantilist rule and from absolutistic regimentation'. It allowed the principle of publicity to be directed against established authorities (Habermas 1989, 56). However, by the end of the eighteenth century, significant tensions were appearing, leading to what he calls the structural transformation of the public sphere. First, the idea of an homogenous reading public was strained by the increasing gulf between a minority group of educated readers and a larger group which began to show their different tastes in the literary market place. Even in Germany – where the middle-class readership was comparatively smaller than in Britain – Schiller, Schlegel, and Goethe sometimes referred to this audience as 'rabble', and addressed their own writing to a small, refined readership. Wilhelm von Humboldt spoke of good writers and their readers as members of an exclusive Masonic lodge. One German visitor to England saw the contrast, exclaiming that the major authors there were 'in all hands, and read by all the people' (quoted in Ward 1974, 128–9). Wordsworth and other English Romantics were trying to reach more of this larger audience (Hohendahl 1982, 55). Second, by the middle of the nineteenth century, democratic reform was beginning to undermine the concept of a uniform public constituted by shared political and social values. Now the 'principle of publicity' was no longer seen as an agency of social critique; instead the issue was the expansion of the public by extensions of the franchise. Thus J. S. Mill, in *On liberty*, regarded 'public opinion' not as the guarantor of rational debate, but as a threat to individual thought and freedom (Habermas 1989, 133).

The Victorian periodicals were one of the last bastions of the public sphere inaugurated in the eighteenth century. There is no doubt that they were the dominant forum for cultural debate amongst the educated upper middle classes and the governing elite. As T. W. Heyck has suggested, in the period before the 1867 Reform Bill they spoke to the small electorate of Britain and to all the 'people who counted in decision-making' (Heyck 1982, 36–7; for further analysis, see Shattock 1989). But even as they were being founded in the first two decades of the century, the conditions that supported the earlier public sphere had begun to collapse. What symptoms of this are discernible in the discourse supported by the major quarterlies – the medium of Young's common intellectual context? At least three can be discerned.

First, it is significant that each of the major reviews carried a political or ideological flag. As is well known, the *Quarterly Review* was founded in 1809 as a Tory counter to the perceived Whig platform of the *Edinburgh Review*, which began in 1802. *Blackwood's Magazine* (1817), the *Westminster Review* (1824), *Fraser's Magazine* (1830), the *North British Review* (1844), and others were connected in some way with different positions in the political and cultural spectrum. The significant point is not the consistency with which these allegiances were manifested in the pages of the journals, but the fact that the sphere of educated discourse was now a site of contest rather than consensus. The editors of the *Edinburgh Review* saw themselves as breaking decisively from eighteenth-century journals whose policy they characterized as 'decent feebleness' (H. Cockburn 1874, 124). Yet this was the kind of polite conversation among men of property, conducted in the *Spectator*, the *Tatler*, and the *Gentleman's Magazine* – the media of Habermas's bourgeoisie public sphere when it represented the voice of reason and civility against absolutist intrusions. By the early nineteenth century the leading journals stood for different responses to a new political environment marked by industrialism and working-class agitation.

Second, and as a consequence of the changing social order, the periodical literature of the educated classes was not the only forum for political and cultural debate. Noting the kind of popular culture documented by E. P. Thompson, Eagleton suggests there was a 'counter-public sphere' in the radical press, in Corresponding Societies, in Cobbett's *Political register* (Eagleton 1984, 36).

Third, even though they were read by an educated, upper-middle-class audience, the major periodicals relied on their reviewers to write in a general, synthetic style. Indeed, given the range of subjects covered by the most frequent contributors, such a style was necessary. But there were signs that this policy of generally readable articles on a wide range of subjects was being strained by the increasing specialization of knowledge in many areas. As early as 1810 Francis Jeffrey, one of the founders of the *Edinburgh Review*, warned that the attempt to keep abreast of all subjects might mean that profound knowledge would be replaced by superficial information on a variety of topics: 'Various and superficial knowledge is now not only so common, that the want of it is felt as a disgrace; but the facilities of acquiring it are so great that it is scarcely possible to defend ourselves against its intrusion' (Jeffrey 1810, 168–9).

By the 1830s it was clear that there was no longer a single educated readership. When Sir William Hamilton applied for the Chair of Logic at Edinburgh in 1836 his opponents charged that his philosophical writings were obscure. Two articles in the *Edinburgh Review* were cited as being 'incomprehensible by ordinary readers'. Hamilton's reply was that the new editor, Macvey Napier, stressed that 'the scientific character of the Journal' meant that it could not overlook European works such as those of Victor Cousin, which might require difficult articles. Hamilton saw no reason why a metaphysician, any more than a mathematician or philologer, should aim his review at the comprehension of 'the general reader', and noted that the 'small philosophical public for whom it was intended, did not find it obscure' (Hamilton 1836, 1–2). As editor of the *Encyclopaedia Britannica*, which was aimed at the middle classes rather than at the growing market for cheaper dictionaries and collections, Napier distinguished two audiences. Writing in 1827 to the new owners, he urged that the seventh edition be extended to twenty-four volumes so that it could serve both general readers and 'men of science'. He sought to include 'miscellaneous matter more particularly adapted to the wants and taste of ordinary readers', without treating significant subjects 'in a way too curt and superficial to satisfy those of a higher class' (Napier 1879, 53). Although the periodicals could avoid the technical details expected in an encyclopaedia, there was now a problem of speaking to both experts and general readers. Coleridge recognized this in 1830 in his *Constitution of church and state* where he warned of the danger of 'plebification'. That he did this while advocating the concept of a 'clerisy' – a body of theologians, scholars and men of science charged with cultivating knowledge – suggests that this notion was an omen of the disintegration of the public sphere. In contrast with the eighteenth century, general discourse among educated people now required a special group to hold it together (Coleridge 1972, 53).

SCIENCE IN THE PUBLIC SPHERE

Now that the tensions present in the forum occupied by the major quarterlies have been outlined, it is time to reconsider the place of science in them. There is no reason to disagree with Cannon and Young on the importance of scientific topics in such reviews, but the previous discussion invites a widening of their analysis. Rather than being simply a component of a common intellectual discourse, the presence of science exacerbated the tensions already mentioned.

Here again some of Habermas's views are helpful as a point of orientation. In work that followed his inquiry into the appearance and demise of the classic public sphere, he discussed what he called the 'scientization of politics': namely, the tendency in the twentieth century for technical knowledge and methods to displace debates about values and directions. One of his proposals was that a modern public sphere had to be reconstituted so that in a rational society 'science and technology are mediated with the conduct of life through the minds of its citizens' (Habermas 1970, 74; Hohendahl 1982, 269–71). This issue has become a serious topic of debate in areas such as science policy, technology education, and environmental politics, where the gulf between the momentum of scientific research and the democratic process that funds it has been acknowledged.

With regard to the first half of the nineteenth century it would be anachronistic to frame the issue in this way, because scientific institutions were less autonomous and less well funded by the state, and the gap between science and its public was less severe. Men of science had to justify their activities and their cultural impact in a more direct dialogue with the lay public. One of the virtues of Young's concept of a common intellectual context is that it seeks to describe the framework in which this took place. Science, in his view, was the sustaining force behind a wide-ranging debate about the question of man's place in nature – one that involved scientific, moral, and religious concerns and did not disintegrate until the late nineteenth century. But by looking at the case of science in the context of a public discourse under stress, it is possible to argue that science was one of the major agents in its fragmentation, and that signs of this can be found much earlier than Young or Cannon suggest. Returning to the points made above about the Victorian periodicals – the political strains in the public sphere, the emergence of an alternative medium of debate, and the conflict between specialist and general discussion – we find that science was both affected by them and possibly intensified the tensions they signalled.

(1) Morrell and Thackray have shown how the leaders of the British Association cultivated a non-controversial image for science, judging that the already weak position of science would be worsened if embroiled in political and religious issues. But this made the periodicals rather dangerous territory since, as we have seen, they were intrinsically political. In some cases, this carried through to subjects such as literature: Jeffrey's attack on romantic poetry as a danger to the social order is a well-known example. Politically slanted reviews of works of science were less frequent, but it is clear that the nature of the scientific enterprise was an issue that did become linked with the social and political issues of the day. In the *Edinburgh Review*, Henry Brougham made science a definite part of his campaign for useful knowledge as a recipe for deflecting working-class discontent. Brewster used the pages of both the *Edinburgh Review* and the *Quarterly Review* to promote a vision of science as an agent of technical application and social reform. The books he reviewed – obvious polemic tracts such as Babbage's, treatises on natural theology, or various works on science – all seemed to deliver this message. Both these prolific reviewers advanced opinions on science that conflicted

with those held by other commentators such as Whewell or Herschel or Sedgwick. In chapter 4 we shall see that the periodicals were an indispensable forum for science; but it is important to note here that scientific topics were not immune to the political cleavages that the periodicals now expressed. By the time of the First Reform Bill in 1832 the public sphere no longer manifested the self-contained discourse among citizens agreed on fundamentals of politics and taste, as Habermas claims it did in the eighteenth century. Instead, it was an area in which rival groups fought for wider support. Babbage had certainly perceived the new situation when he wrote to Herschel at the time of his attack on the Royal Society: 'with the aid of public opinion I will make them writhe if they do not reform' (Hall 1984, 49).

(2) Outside the major periodicals there were even more divergent notions of science, its political affiliations and social purposes. The work of Berman on the involvement of utilitarians in the Royal Institution, of Cooter on phrenology, and of Desmond on artisan and radical evolutionary speculation, maps out another forum in which science was defined and deployed (Berman 1978; Cooter 1984; Desmond 1989). Inkster's account of the Seditious Meetings Act of 1817 suggests that at that time science may have been seen as a threat to established public culture (Inkster 1979). Although this territory is outside the scope of this book, its presence is felt in so far as Whewell and other leading commentators on science were compelled to deal with it, often by guarding their ideas of science against its intrusion.

(3) Young has rightly urged that in early Victorian debates the division between science and what we would now call the arts was not a rigid one (R. M. Young 1985, 132). But specialization was nevertheless an issue closely connected with science and its involvement in the public sphere in the early decades of the century. Jeffrey is again a good monitor of the problem of maintaining a general discourse on culture in the face of growing intellectual specialization. Reflecting on the implications of this for contemporary ideas about the appropriate body of 'general knowledge' he complained that: 'Now-a-days ... a man can scarcely pass warrant in the informed circles of society, without knowing something of political economy, chemistry, mineralogy, geology and etymology' (Jeffrey 1810, 168). Significantly, scientific disciplines were prominent in this list of subjects exploding the limits of feasible 'general knowledge' and, of

course, the capacities of the non-expert reviewer. The rapid progress of science, especially in France, also led some men of science to criticize the standard of scientific discussion in Britain. Herschel may have been commenting on the limits of this public forum when he included a controversial footnote in his article on 'Sound' in the *Encyclopaedia Metropolitana*, complaining that there was no expert audience for original science. He contrasted the standard of Continental scientific journals with the 'crude and undigested scientific matter which suffices . . . for the monthly and quarterly amusement of our own countrymen' (Herschel *c.* 1830, 810). By the time of the controversy over the anonymous *Vestiges of the natural history of creation* in 1844, the needs and interests of the lay public and the specialists were in conflict. In the earlier decades the periodicals hosted a conversation between leading men of science to which the public were invited as participants; in 1844 they were a medium in which specialists announced their dismay at public taste in scientific literature (see Yeo 1984 and ch. 4 below).

In spite of the tensions associated with science in this fragmenting public sphere, science had to be conducted and defended, to a significant extent, in the public domain. Under these pressures the appeal of a flexible, anodyne natural theology is understandable; but this in itself was insufficient as a way of defining and defending science. Leading men of science had to say more than this and they had to say it in a public arena contested by various groups with different agendas and different uses for science. When they did speak they disagreed among themselves on natural theology itself, on method, on the organization of research, on the history of science and on other issues. Consequently, the metascientific debate this generated was not only about the nature of science, but concerned the best way of speaking about it in various forums.

These are just some of the issues that made Whewell's bid for the role of metascientist a timely one. But in the 1830s, as Whewell began to enter the public forum, this was not a well-established intellectual function. Even after his major works, it is apparent that some reviewers, especially non-scientists, found them difficult to categorize: the *Philosophy* suggested comparisons with Kant, but its focus on the physical sciences was seen as unusual. One suggested that Whewell offered 'a bird's eye view' of the history of science ([Ulrici] 1847, 4); another likened his work to 'criticism of poetry', but noted that 'histories and philosophies of science' were peculiar to the present age:

only recently had inductive science itself become 'a subject of induction' ([Butler] 1841, 194–5). The next chapter examines the way in which Whewell came to realize that in order to define science he had to define a role for himself as its critic.

CHAPTER 3

Metascience as a vocation

> ... the philosopher is not an artificer in the field of reason, but himself the lawgiver of human reason.
>
> Kant 1933, 658

> If this course educate a man for anything, it educates him to be a judge of philosophical systems; – an office which so few Englishmen will ever have to fill.
>
> Whewell 1838a, 49

Unlike Merz in 1896, Herbert Spencer, writing some fifty years earlier, believed that the definition of science was by no means a straightforward or finished matter. Indeed, in an essay review on 'The genesis of science', he questioned many of the conclusions of the two major scientific commentators of the age: Auguste Comte and William Whewell. To begin with, he claimed that both endorsed the dominant view that 'scientific knowledge somehow differs in nature from ordinary knowledge'. This absolute contrast, in his opinion, removed the possibility of understanding the growth of science as a process of development on evolutionary lines. While welcoming Comte's efforts Spencer rejected his notion of a 'serial' progress of the sciences: scientific advance was 'as much from the special to the general as from the general to the special' (Spencer 1854, 108, 161). Furthermore, he argued that although Comte recognized a 'common origin' of the sciences, he wrongly presented the subsequent divisions as separate, overlooking the ways in which they interacted, forming new syntheses. Here Spencer seemed to find support in Whewell's *History*, suggesting that this 'mutual influence of the sciences has been quite independent of any supposed hierarchical order', referring to Whewell's account of how the latest progress of chemistry depended on physiology to produce galvanism. But, on the other hand, Spencer's attack on classifications of science as artificial, and his

49

hostility to distinctions between science and art, or science and common sense, must be read as antagonism to Whewell's project (Spencer 1854, 138, 152–9).

These rather intense examples show that Spencer had much to argue against; and this indicates that the nature of science and its relationship with other forms of knowledge was still very much a contentious issue. This observation fits with a recognition of the considerable philosophical reflection on science during the first half of the nineteenth century – a discourse not yet divorced from the actual practice of science. As Larry Laudan has emphasized, some of the most significant commentaries on the method of science in this period came from practising scientists – Herschel, Brewster, Whewell in Britain; LeSage, Bernard, and Poincaré in France (Laudan 1981, 7, 13). This list could be extended by adding figures such as Powell and De Morgan.

In the previous chapter I argued that the social and intellectual conditions of science in the early Victorian period supported a metascientific discourse. This was conducted through a variety of media and was not confined to the formal philosophical texts that later came to be the main site for such commentary. Men of science in early-nineteenth-century Britain knew that they had to secure a place for science in the public forums and not just in specialized scientific societies. This required the affirmation of the value of the scientific enterprise as well as defence of particular theories and discoveries. In this sense, most men of science were also engaged in metascientific discussion. But by 1830 Comte was prescribing a specialized role for this activity.

In his *Introduction to positive philosophy*, Comte remarked on the potential danger of excessive specialization in science and warned of the intellectual isolation of particular pieces of research from wider generalizations. Although he regarded this as a product of the division of scientific labour, Comte suggested that just one further division could resolve the problem:

All that is necessary is to create one more great speciality, consisting in the study of general scientific traits. We need a new class of properly trained scientists who, instead of devoting themselves to the special study of any particular branch of science, shall employ themselves solely in the consideration of the different positive sciences in their present state. (Comte 1970, 17)

Comte obviously had a very specific view of this task because he proposed a Chair in History of Science to Guizot and recommended himself as the only suitable candidate (Lepenies 1988, 21). At this time, but from a different perspective, Coleridge also regarded the specialization of science as a problem; however, his solution was not the creation of a special function within the scientific community, but the revival of the ancient notion of philosopher – a term he refused to the men assembled at York for the first meeting of the British Association. For Coleridge, philosophy was second only to theology in importance and it could serve as 'the Supplement of Science' (Levere 1989, 86 and 1981, 73; also Corrigan 1980, 403).

Neither the French nor the British scientific community formally adopted such an explicit notion of a metascientific class. It is possible that in France, where science was more securely placed in its connection with a State-administered education system, there was a more obvious role for Comte's comparative comments on the sciences, the boundaries between them, and the best order of study. Nevertheless, his diagnosis of an increasingly specialized scientific enterprise found independent notice in the discussions at the time of the foundation of the British Association. Most of the leading members accepted the need for a careful survey of the present state of natural knowledge – a 'map of science' as Herschel described it to the first secretary, William Vernon Harcourt (Herschel to Harcourt, 5 September 1831, in Morrell and Thackray 1984, 55). It was Whewell who responded to this most deliberately by suggesting the commissioning of reports on various sciences (Whewell to Harcourt, 1 September 1831, in Todhunter 1976, II, 126–30). In doing so he was following up his earlier suggestion to Peter Mark Roget, as Secretary of the Royal Society, that it should encourage critical surveys of various scientific areas, such as those by Cuvier, Fourier, and Berzelius (Whewell to Roget, 22 March 1831, RS Domestic MS. 1, no. 30). Although Whewell spoke of these as being 'somewhat popular in style', he did not envisage them as directed beyond a scientific audience. Eventually, the reports for the British Association were written by experts in particular fields, but Whewell's own, on mineralogy, and on electricity and magnetism, were more than summaries of present achievements: they were exercises in comparative historical analysis, assuming a general model of scientific progress against which developments in particular disciplines could be assessed (see Whewell 1832b and 1835).

There is little doubt that Whewell can be seen as the person in

Britain who best answered, or even anticipated, Comte's prescription of this role. Unlike his colleagues in the scientific world he was seeking to offer commentary and criticism, rather than practising science itself. Such a role was quite distinct from the way in which leading members of the scientific community might be seen as members of a Coleridgean clerisy, together with other scholars in law, medicine, architecture, music, and theology (Coleridge 1972, 36; also ch. 8 below). If anything, Whewell's activity was closer to Coleridge's idea of a philosopher standing above the various professions and bodies of knowledge. Murchison seemed to recognize this in a negative sense when he spoke of Whewell in 1843 as a 'high priest of science' aloof from the 'men of science' (cited in Morrell and Thackray 1981, 430).

By this time Whewell had a definite sense of his relation to the scientific community and his place in the wider intellectual community. But he did not always possess this, and over the previous two decades he was concerned with the nature of his vocation. In a substantial biographical essay, Leslie Stephen later reflected the puzzle Whewell had set both his contemporaries and himself: as Stephen put it, 'Whewell began as a man of science' but then 'scarcely became a philosopher' (Stephen 1885–90, 20, 1,371). This chapter considers how Whewell understood the metascientific role he adopted over this period. This will require an appreciation of why Whewell was doubtful about his status as a man of science, and why an alternative conception of his activities was difficult to achieve.

LOOKER-ON ON SCIENCE

As early as 1818 Whewell seems to have cast himself in the role of an observer rather than a maker of science. Writing to Herschel about recent work in optics, he spoke of it as a rich field of potential discoveries and assessed the places of Brewster, Biot, and Herschel in 'the race of discovery'. He referred to the rapidly accumulating facts about the properties of light and had no doubts that 'some of the general laws which are enveloped in them' would fall to Herschel. But Whewell situated himself outside this activity, saying that he was one of those 'lookers on, who, not making a single experiment to further the progress of science, employ ourselves with twisting the results of other people into all possible speculations mathematical, physical and metaphysical' (Whewell to Herschel, 1 November 1818, in Todhunter 1876, II, 28–9).

If this is taken as an early negative self-assessment of his scientific potential it is curious that Whewell should have made it. As second wrangler in 1816 he had the credentials for major participation in science, and might have been expected to use the security of his Fellowship, achieved in 1817, as a basis for such activity. The path he did take, from assistant mathematics tutor in 1818, head tutor at Trinity in 1823 and Professor of Mineralogy in 1828, was compatible with the kind of scientific vocation pursued by colleagues such as Adam Sedgwick in geology and George Airy in astronomy. In 1823 he told Herschel that, when admitted to the Royal Society, 'I intended, if possible, to avoid belonging to the class of absolutely inactive members, and I have since been on the look out to find among the speculations that come in my way some one which might possibly be worth presenting to it.' In response to a request for comments on a paper about mathematical aspects of crystallography, Herschel affirmed that it was 'fit for the transactions of any Society in the world' (Whewell to Herschel, 15 October 1823; Herschel to Whewell, 15 October 1823, HP, vol. 18, nos. 163–4).

Recent accounts of Whewell's practical scientific work by Becher and Ruse remind us of his considerable presence in a number of fields. While his mechanics texts, starting from 1819, reflected pedagogic rather than research interests, Whewell was involved in geological expeditions with Sedgwick from 1821 and sought the most advanced instruction in mineralogy and crystallography in Berlin, Freiburg, and Vienna in 1825. In June 1826 he set off to Cornwall with Airy to spend several weeks in a mine shaft experimenting on the mean density of the earth. The plan was to compare the effect of gravity on invariable pendulums at the surface and at a depth of 1200 feet (Todhunter 1876, 1, 37–40; Stair-Douglas 1881, 101–4). The experiment was not successful and another attempt was made two years later. A description of their efforts was published as *Account of experiments at Dolcoath mine in Cornwall* in 1828. In that year Whewell became Professor of Mineralogy, nominating for the position on the platform of applying mathematics to crystallography and improving classification in mineralogy. He had already published five papers in the area (Whewell 1828b; Becher 1991; Ruse 1991).

Whewell's most significant scientific contribution was his study of tides, recorded in fourteen papers presented to the Royal Society of London from 1833 to 1859. Thus shortly after suggesting to Roget and Harcourt his plans for mapping international scientific research

Whewell was engaged, with John Lubbock (his former student), in a quest to chart the movements of the world's oceans. The aim was to produce a 'global map of cotidal lines', showing the points where 'high water occurs at the same time' (Becher 1991, 13; Morrell and Thackray 1981, 513–17). For this he needed to recruit observers, and as well as drawing on scientific societies and coast guards, he even consulted the *Missionary Magazine* and his sister's contacts with this movement, as a source of information (Whewell to Martha, 13 February 1835, in Stair-Douglas 1881, 171). Although he was not fully satisfied with the results of his work, he was rewarded in 1837 with a Royal Medal from the Royal Society. But two years earlier he maintained this topic was 'for my own personal satisfaction' and that he was not troubled by the judgements of others. It is significant that when telling Herschel of giving 'the rest of my life' to metascientific inquiries, he added that 'I always reserve to myself the *tides*, as a corner of physics which I shall go on and work at' (Whewell to Herschel, 9 April 1836, in Todhunter 1876, II, 235).

These activities were sufficient to allow Whewell's election to the Royal Society in 1820, admission to the Geological Society in 1827, and nomination for its presidency in 1837. His influence during the 1830s on the scientific terminology of Lyell and Faraday was profound, and in the latter case, connected closely with the fundamental concepts at stake (Whewell to Faraday, 25 September 1835, WP, O 15. 47[148] and Schaffer 1991, 226–30). Given all this, we need to be cautious in accepting Whewell's various comments about his low scientific achievement. But we also need to appreciate the perspective from which he made this assessment.

Some of Whewell's remarks refer to the fact that he did not consider himself a major scientific discoverer. His contributions in mineralogy and 'tidology' – as he called it – were important, but neither met his own criteria for truly significant advances in science, and they did not compare with those of the leading men of science he counted among his friends. In the case of mineralogy he saw it as the fate of being in the wrong scientific field. Thus in 1827 he explained to Herschel that he was happy with the reforms he had suggested for the subject, but grieved 'at the small chance there is of my ever making those original discoveries and advances in the science which give a man the right and power to regulate its external clothing' (Whewell to Herschel, 23 November 1827, in Todhunter 1876, II, 85–6). In the case of the research on tides, he felt inadequate in not being able to push beyond careful observation to an advance in 'hydrodynamical theory', confessing to Forbes that being 'long unused to the heavier weapons of analysis, I do not dare to attack so formidable a problem' (Whewell to Forbes, 2 April 1838, in Todhunter 1876, II, 269). Herschel seemed to be compensating for his friend's self-perception when he later remarked of Whewell's textbooks on mechanics, that 'their sterling value will secure them an estimation superior even to that of many original discoveries' (Herschel 1841, 217; contrast Todhunter 1876, I, 20 on the lack of 'stability and permanence' in these due to revisions).

Whewell's scientific work gave him respected membership of the leading scientific societies, in which he was also acknowledged as a powerful and politically astute organizer; his opinion on nomenclature was valued by Lyell and Faraday; and he was sought after as a scientific reviewer. In addition, he assumed major roles in College and university administration and educational reform. Yet in spite of his declared empathy with Trinity and its traditions, we know by Whewell's own admission – in private letters – that he sought a grander role than that of a university academic. He did not lack employment; but he still sought to clarify a vocation.

When trying to refuse Murchison's efforts to draft him into the Presidency of the British Association in 1840, Whewell devalued his scientific achievements, saying that 'there is nothing of such a stamp, in what I have attempted, as entitles me to be considered an eminent man of science' (Whewell to Murchison, 18 September 1840, in Todhunter 1876, II, 286). This was not simply an indication of a recent shift of interest from science to moral philosophy, as Todhunter suggested; rather, it was part of a stance toward science evident, as we have seen, as early as 1818, thus predating his mature philosophy of science. In 1826 when he contemplated the vacant Lucasian Chair in mathematics, he told Jones that he would use it to 'make very grand lectures on the principles of induction' (Whewell to Jones, 13 October 1826, in Todhunter 1876, II, 72; cited in Becher 1991, 16).

Rather than seeing Whewell's lack of scientific achievements as the problem here, it is worth focusing on the nature of his major project – the philosophy of induction. There is a problem about the precise nature of this activity and its place in the recognized social and intellectual functions of the time. We can appreciate this by underlying the fact that *science* was the object of Whewell's critical attention. There was a recognized 'critical' role in relation to poetry, fictional literature, and drama, and works of history, biography, and travel. But the case of science was different, partly because (as indicated in the last chapter) it did not offer an established vocational path. We need only look at the cases of two of the most influential contemporary men of science – Georges Cuvier, who died in 1832, and Charles Darwin, who began to contemplate a life in science at this time. The work of Dorinda Outram and James Moore shows, in different ways, how both these figures struggled with the idea of a vocation in science. Significantly, both had to do this by reflecting on the relationship of science to other spheres of activity: in Cuvier's case, a real life in politics and State service; in Darwin's, an imagined life in the Church as the 'vicar of Down' (Outram 1984; Moore 1985).[1]

Whewell's case was more complex because he sought to define a role as critic, adjudicator, and legislator of science without being a major scientific practitioner or discoverer. This made his situation different from that of Herschel or Brewster, who could comment on issues of method or theory, often in relation to their own original investigations. Whewell could do this too, and we should not forget the work he did on mineralogy and tides; but he seemed to doubt whether this was an adequate basis from which to speak. The other problem was that Whewell's metascientific concerns extended beyond the mathematico-physical science, where he had competency, to the full range of organic disciplines.

There have been two ways of conceiving what Whewell did and what he was. He has been seen as an omniscient polymath devouring the full range of physical sciences, together with the classical and literary canon which usually contented other universalists such as T. B. Macaulay. This image is not sufficiently precise. On the other hand, there is the account of Whewell as the father of modern history and philosophy of science. This is too specific and, unless carefully qualified, anachronistic as well. The remainder of this chapter will attempt to check these two extremes by asking how Whewell came to adopt the role of a metascientist.

WHEWELL AS OMNISCIENT

In her diary for 4 September 1859 Caroline Fox reported Whewell's admission that he was still trying 'to live down Sydney Smith's quiz

[1] More generally, on the problem of the idea of the 'scholar' at a time when the ideal of the 'gentleman' was dominant in English culture, see Shapin 1991. The vocation of a 'man of science' was often defined by reference to established roles, such as that of clergyman. Chapter 5 deals with the moral expectations linked with the idea of science.

about Science being his forte, and Omniscience his foible' (Monk 1972, 228). While this notion of universal knowledge has always figured in discussions of Whewell it is seldom interrogated. What function did it have for Whewell's perception of himself and his intellectual direction? What was its contemporary meaning?

There is ample evidence that Whewell regarded omniscience as an attractive goal. His earliest letters from Trinity contain frequent references to the 'Demon of universal knowledge' and a confession of 'certain indefinite desires to approximate to something like omniscience' (Whewell to Morland, 10 August 1815, in Todhunter 1876, II, 8, 10). One way of approaching this theme is to view it as a means of expressing interests beyond the narrow mathematical confines of the Tripos. It is significant that he made these remarks in letters to George Morland (a teacher at Lancaster Grammar school) when expressing an interest in metaphysics – by which he meant the writings of Berkeley, Reid, and Stewart. Whewell claimed that this was a 'very useful study', but he must have known that it was not likely to determine his rank in the Tripos examination. When he won the Chancellor's medal for an English poem on Boadicea in 1814 he explained to his family that he was not neglecting his studies. Joseph Rowley, his former master, was not impressed, saying that:

he has gone and got the Chancellor's gold medal for some trumpery poem, 'Boadicea', or something of that kind, when he ought to have been sticking to his mathematics. I give him up now. Taking after his poor mother, I suppose. (Stair-Douglas 1881, 4)

The omniscience theme might also indicate the excited and slightly confused reaction of a youth catapulted from an artisan family in Lancaster to the centre of academic learning. Whewell told Morland that he was 'seized with an inconceivable desire to read all manner of books at once', and was impressed by the realization that he had 'the means of getting at almost every book that ever was written' (Whewell to Morland, 3 January 1815 and 10 August 1815, in Todhunter 1876, I, 5 and II, 8).

But the context in which he made these remarks is revealing. Having visited London for the first time in 1815, Whewell admitted to his sisters that he had only seen the city from 'the *outside*' because, not knowing anyone there, he could not 'see anything of its society' (Whewell to his sister, 14 April 1815, WP, Add. MS. a. 301[2]). He was still pondering this experience four months later, and confessed to

Morland that he:

began to be tired of seeing a world where every body but myself seemed to have something to do, where society seemed made up of sets without leaving any vacancy which I could fill, and accordingly I left them, knowing that they would mind their business just as well if I were in my dressing gown at Cambridge. I very soon by the help of philosophy overcame the mortification of finding that I was not ... a necessary part of the scheme of things. (Whewell to Morland, 10 August 1815, in Todhunter 1876, II, 7)

Juxtaposed with this is Whewell's realization that he was seen as exceptional precisely because he had intellectual ability without social status. Thus he acknowledged that it was 'more than ever necessary to read in order to come as near as may be to the expectations that my friends entertain' (Whewell to Morland, 15 January 1814, in Todhunter, 1876, II, 6). There may be a strong sense in which Whewell's connection with omniscience was part of a process of personal and social legitimation.

This may be so; but it is worth noting that even at the time of Whewell's early university days, the ideal of omniscience was seen as untenable. In his novel *Headlong Hall* of 1816 Thomas Love Peacock satirized the prospect of an omniscient individual when he described Mr Panscope (probably Coleridge) as:

the chemical, botanical, geological, astronomical, mathematical, metaphysical, meteorological, anatomical, physiological, galvanistical, musical, pictorial, bibliographical, critical philosopher, who had run through the whole circle of the sciences, and understood them all equally well. (Peacock 1893, 68)

More specifically, the ideal of a circle of science or knowledge, to which Whewell referred in one of his letters, was being questioned in encyclopaedias – the texts which had previously embodied this concept. In his Preliminary Dissertation to the sixth edition of the *Encyclopaedia Britannica*, Stewart doubted the value of systematic classification of the sciences, suggesting that the complex and changing state of modern knowledge made it impossible for a single person to grasp the relationships between the various disciplines (Yeo 1991b, 30–2). The first volume of this appeared in 1815, the year in which Whewell was confessing his dreams of universal knowledge to Morland.

It is therefore worth looking more carefully at Whewell's reputation for omniscience. Here it is useful to compare him with Macaulay, since they were both seen, along with Coleridge, as the greatest talkers of the day – a capacity sustained by extensive and various knowledge. After a dinner in 1832 Romilly noted: 'Much brilliant conversation from Macaulay and Whewell: – no politics at all' (6 October 1832, in Bury 1967). But was Macaulay's omniscience similar to Whewell's? To begin with, Macaulay admitted his lack of mathematical and scientific knowledge, freely confessing to Whewell that this was partly due to 'foolish idleness at Cambridge' where he had 'gulfed', that is failed to gain a place in the Senate exams. Earlier he had admired Herschel's *Discourse*, expressing surprise at the vigour of its 'style' and remarking to Whewell that 'Herschel must be a man of letters as well as a man of science' (Macaulay to Whewell 3 July 1838 and 5 February 1831, WP, Add. MS. a. 209[159] [and] [150]).

This suggests two observations: first, that it was possible to be seen as polymathic, like Macaulay, without showing proficiency in science; and second, that extensive knowledge of science was thought to rule out equal grasp of other subjects. Omniscience now had limits. Whewell was seen as amazing because he knew science but also the classics and modern literature; he seemed to be transcending such limits. Some years later when Macaulay received a volume of English hexameters by Whewell and Herschel, he replied that 'you have satisfied me ... that a genius for poetry and a genius for the severer sciences are perfectly compatible'.[2]

In fact, though, Whewell was aware that he did not possess a comprehensive knowledge of the full range of sciences. He relied on Richard Owen for detailed points on physiology and comparative anatomy in the *History*; and confessed to James Garth Marshall (his brother-in-law) that 'the only two branches of physical science of which I know anything are crystallography and the undulatory theory of Light' (Whewell to Marshall, 19 August 1854, in Stair-Douglas 1881, 436).

Rather than seeing his quest for a metascientific role as a simple extension of early desires for omniscience, it is more useful to interpret it as a response to the expansion and diversification of science that rendered omniscience of the eighteenth-century kind impossible. When Whewell and Jones were sketching their plans for a broad

---

[2] Macaulay to Whewell, 25 March 1849, WP, Add. MS. a. 209[156]; see also their correspondence on the Platonic Dialogues: WP, Add. MS. a. 209[164]. Referring to Brougham, Macaulay said: 'I have not the Chancellor's Encyclopaedic mind. He is indeed a kind of Semi Solomon. He *half knows* everything' (Macaulay to Napier, 17 December 1830, BL, Napier MSS. 34, 614, f. 460). See also Robinson 1872, II, 226 for a recognition of Whewell's pursuits in 1840 as 'very multifarious'; also Herschel 1867–8.

study of inductive philosophy there were warnings about the danger of seeing this as an encyclopaedic venture. For example, in 1824 the Scottish philosopher Alexander Blair denied that individual minds could grasp a unity across an encyclopaedic range of subjects. Furthermore, he claimed that knowledge advanced by specialist research, not by the pursuit of unifying ideals (Blair 1824, 26, 32). In the same year Thomas DeQuincey admitted that 'a few Leibnitzes in every age' would help other people to grasp some of the connections between different aspects of knowledge, but he argued that there was no point in trying to recover the more comprehensive scope of past thinkers, or the rounded culture of classical Greece. The modern period had lost these things, but the 'subdivision' of intellectual work had produced results that were unattainable in earlier times (De-Quincey 1824, 26–7).

It is significant that both these writers mentioned science as one of the main factors in the recent demise of the omniscient individual. Indeed, S. T. Coleridge, one of the last examples of this category, had underlined this point in 1817 in his 'Preliminary Discourse' to the *Encyclopaedia Metropolitana*. Unlike Stewart, Coleridge believed that a map of the sciences was the essential basis for the organization of such a work (Yeo 1991b, 33–4). His classification distinguished between pure, mixed, and applied sciences. Since only the first of these could offer certain, demonstrative knowledge from first principles, the other branches of science were to some extent unstable. The relevance here is that much scientific knowledge was constantly changing, especially in subjects like magnetism, electricity, galvanism, and chemistry which at present, in his estimate, all lacked clarity in their 'first idea'. In contrast to the more exact sciences where present knowledge was 'not likely to be superseded by any new discoveries', the theories of such disciplines were usually imperfect because they were 'necessarily progressive': it was always possible that some new discovery might alter the theory. Thus geology and chemistry were 'constantly enlarging their boundaries and changing even some of their elementary principles' (Coleridge 1817, xiii, 35–40). This feature of the 'progressive' sciences made it difficult both to maintain an encyclopaedia and to be an omniscient person.

Thus although Whewell could be seen as 'a sort of Cambridge Leibnitz', he was not seeking to revive a past ideal (Butler 1841, 197). Whewell appreciated the situation Coleridge described, and recognized that he depended on experts for access to the latest scientific knowledge. In 1828 he told Herschel he knew nothing of 'Sound' and was waiting to read his article on it in the *Encyclopaedia Metropolitana* (Whewell to Herschel, 14 October 1828, in Todhunter 1876, II, 95). Whewell realized that because no individual could know the entire circle of sciences, it was essential to compare and classify the different branches, to look for analogies between them.

WHEWELL AS PHILOSOPHER OF SCIENCE

Although shortly after his death some writers referred to Whewell as the 'first historian and philosopher' of the inductive sciences, it is only twentieth-century commentators who have focused exclusively on this dimension (Carlisle 1882, 144). Whewell's friends and enemies also knew him as a theological apologist, an active member of some twenty scientific and scholarly institutions, a translator of Greek and German verse, a strong campaigner on educational theory and practice, twice Vice-Chancellor of Cambridge. Undoubtedly, his major works can now be seen as foundational texts in the subjects of history and philosophy of science, but it is important not to abstract these from his other concerns. Some accounts of how Whewell came to be engaged in philosophy of science have done this by pushing his theological, moral, and educational preoccupations to the periphery, or by constructing him as an English Kantian irritated by the prevailing empiricism. As Fisch has recently argued, there have been few serious attempts to delineate the path that led to Whewell's major works. The reconstruction Fisch offers is valuable because it links Whewell's mature works with one of his earlier engagements as an author of textbooks on mechanics, and hence with his later educational concerns. He published *An elementary treatise on mechanics* in 1819 and *A treatise on dynamics* in 1823; these were followed by other editions and further textbooks – the *Mechanical Euclid* of 1837 being the last. It was through this activity that Whewell 'first encountered science as an *object of study*', well before he commenced the notebooks that led to the writing of his major works (Fisch 1991a, 41). Whewell did not publish on general scientific issues until 1831, when he reviewed Herschel's *Discourse*, but on this reading the foundations of his later metascientific project were established through the textbooks on mechanics (Fisch 1991a, 8, 14–15). It may also be that an academic reputation in mathematics allowed Whewell to assume this higher vantage point.

However, it is important to remember that in numerous places Whewell called his project a 'philosophy of knowledge', which sought the reform of philosophy. In the 1820s, Whewell and Jones imagined a study of 'inductive philosophy' that went beyond any particular science. Mechanics, mineralogy, meteorology, and political economy were all grist to their mill, but when Whewell referred to 'that higher philosophy of yours which legislates for sciences', he expressed their dominant passion. There was more at stake here than sound reasoning: Jones believed that 'systems palpably mischievous and immoral' had attracted hearers, due to an inadequate public understanding of induction and its relevance to moral and social topics (Whewell to Jones, October 1825, in Todhunter 1876, II, 61; Jones to Whewell, 27 September 1827, WP, Add. MS. c. 52[15]). Their idea of a journal covering 'moral philosophy, political economy, and science' was part of this reforming programme. By 1830 the prospect of contributing to this cause was, in their view, at least as noble as scientific discovery. Jones hoped that his friend would have:

a niche in literary history near Bacon and Locke and far above us all as far as we can see our way yet. I expect however something great from Herschel if he lives and does not let astronomy engulf him at last, about which I have my fears. (Jones to Whewell, November 1831, WP, Add. MS. c. 52[42])

As mentioned in chapter 1, there is an interpretation, instigated by Todhunter, which views Whewell's work after 1840 as the translation of focus from physical to moral science, or the application of his views on method and epistemology to these additional spheres. Yet both interests were already present in his early correspondence with Jones, and chapter 7 will emphasize the interaction between moral and scientific concerns from the start of Whewell's philosophical endeavours. For although Fisch is correct in noting that Whewell did not *publish* any philosophical material on science before 1831, he did begin to confront the moral and intellectual status of science in these early discussions. His dialogue with Jones is important here because this combined consideration of inductive method with moral, social, and metaphysical issues. But neither the image of Whewell as driven towards omniscience, nor that of him as engaged in pursuit of a philosophical account of excellent physical science adequately captures this dimension of his quest for a vocation as a metascientist.

Defining a role

The problem here can be posed by asking one question: what models were available to Whewell as a critic of science? In order to avoid anachronism we need to be clear that in some ways Whewell was striving for a role that did not exist. At this time in Britain, there was nothing like the idea of a specific class of intellectuals of the kind emerging in France and Russia. As various writers have argued, the sense of an oppositional group, characteristic of these other intelligentsia, was absent (Kent 1978, xi–xii; Heyck 1982). In Whewell's most immediate environment – the university – the concept of a lifetime academic tenure as a possible basis for an intellectual career was not fully established until mid-century, well after his major works (Robson 1967, 318; Rothblatt 1981). Until they left for careers in the Church or government, Fellows and tutors were members of an ecclesiastical institution where research was not an expected function. Outside the university, the idea of writers as critics of literature and society began to appear from 1800, especially under the influence of Romanticism, and in a dependent relationship with a reading public (Gross 1969). Was there, then, in the early part of the century, any counterpart in the field of science to the cultural criticism of Wordsworth, Coleridge, or Carlyle?

One possibility is the school of Scottish Common Sense Philosophy. Dugald Stewart, the surviving representative of this Enlightenment tradition, did use his lectures and writings to reflect on the nature of science. In the second volume of his *Elements of the philosophy of the human mind*, published in 1814, Stewart gave an account of inductive logic as outlined by Bacon. His prime focus was not the actual practice of science or its history, but the clarification of terms and concepts carried into Baconian induction from Aristotelian logic – for example, analysis, synthesis, induction. He also defended the 'method of hypothesis' against 'the indiscriminate zeal' of Bacon's disciples (Stewart 1854, III, 307–14; Yeo 1985, 263–66; Corsi 1988, 42–5. Stewart's first volume was published in 1792).

Whewell read Stewart in his early student years when he explored the realm of 'metaphysics' that lay outside the Cambridge curriculum. This was undoubtedly an important exposure to wider philosophical issues but did not mean that there was a ready-made model as Whewell began to conceive his vocation as a critic of science. The comparatively small part of Stewart's writing dealing with physical science was subordinate to what he saw as the primary concern of 'the Philosopher' – namely, the nature of the human mind as a basis for an understanding of social behaviour. He contrasted the knowledge of 'the Philosopher' with the 'sagacity which directs uneducated men', and taught moral philosophy, a foundational subject in Scottish universities, to bring this disciplined reflection to other activities (Stewart 1793, 3; Davie 1964). The absence of this approach at Cambridge meant that the role of philosophical commentator was not easily available to Whewell in his positions of college tutor or Fellow.

The case of Baden Powell, however, is closer to Whewell's. Like Whewell he was a mathematically trained university teacher and not a major scientific researcher. However, unlike Whewell, Powell was brought up in evangelical circles associated with the Clapham sect, which has been seen as one of the formative points of the Victorian intellectual aristocracy. He was sent to Oriel College at Oxford to be trained by Copleston, Whately, and Newman, the so-called Noetic group who cultivated a tradition of monitoring current philosophical ideas. Powell presented his reports on science and its theological implications as a product of this role (Annan 1955, 244; Corsi 1988). In contrast, Whewell did not have such a close link with a tradition of commentary, based on family connections and denominational allegiance, to legitimate his activities. Nor did he have the detailed Scriptural and theological training that Powell employed when writing on scientific questions. Eventually, of course, he had an unrivalled knowledge of the history of science as a basis for his general comments, but this was not present in the early stage of his career, and it was not the equivalent of theology as a vehicle for public debate. Another way of putting this might be to say that Whewell's marginal social background made it more difficult for him to assume the role he was casting for himself.

These examples indicate that the adoption of a critical role in relation to philosophy or science was closely bound to specific contexts – the Scottish curriculum, in Stewart's case; a particular religious tradition, in Powell's. Whewell's Cambridge setting did not provide him with this; he had to make his own rationale for a critical commentary on science, and he did this by drawing on other cultural resources: Romanticism, and the more established idea of intellectual service to Church. His relationship with Hare, Rose, and to a lesser extent, William Rowan Hamilton, was crucial here. The last section of this chapter considers Whewell's engagement with Romanticism, since this provides a way of seeing that, from an early stage, his philosophical interest in science was part of a wider moral and cultural debate. Chapter 7 returns to this theme by showing how Whewell conceived his philosophy of science as part of moral reform.

SCIENCE AND ROMANTICISM

These two subjects are often seen as antithetical, but there were some intriguing connections in the first two decades of the nineteenth century. At this time, Romantic poetry and, to a lesser extent, science were outside the dominant cultural and academic circles; both had to affirm their significance to contemporary society. Romantic poets and men of science were seeking to create an audience for their activities; in some cases they saw themselves as fighting over the same clientele. Humphry Davy's opening discourse at the Royal Institution in 1805 drew criticism from Wordsworth and Coleridge because it appeared to set the social and cultural benefits of science above those of poetry (Siegfried and Dott 1980; Sharrock 1962). They also defined themselves against each other, thus leaving the oppositions which are now familiar, but concealing the interesting dialectic between them. The studies of Knight, Levere, and others have begun to recover these connections, especially in the work of Davy and Coleridge (Knight 1967; Levere 1981; Cunningham and Jardine 1990). Whewell's case is different, in that Romantic notions were not directly involved in either the poetry he composed or the science he produced, but they were arguably significant for the inauguration of his metascientific approach to scientific issues.

When the young J. S. Mill asked 'What is Poetry?' in an article in the *Monthly Repository* of 1833, he found himself talking about science.[3] This was because he relied on the framework established by the English Romantics in which, as Wordsworth insisted, the 'logical opposite' of poetry was not prose, but 'matter of fact or science' (Mill 1833, 344). In the preface to the second edition of *Lyrical Ballads* in 1800 Wordsworth contrasted the 'Man of Science' with the Poet. The

---

[3] Mill wrote two articles on this topic in 1833: 'What is poetry?' and 'The two kinds of poetry' – both in the *Monthly Repository*. These were combined in a version later reprinted in Mill (1859). See Lepenies 1988, ch. 3, on Mill as cultural critic. The *Quarterly* and *Edinburgh* reviews were not favourable to Romantic poetry; the former often defended Augustan poetics, and in the latter Jeffrey attacked Wordsworth. See Sullivan 1983, 11, 359, 141 and Gross 1969, ch. 1.

first, he wrote, 'seeks truth as a remote and unknown benefactor; he cherishes and loves it in his solitude'; the second sings 'a song in which all human beings join with him, rejoices in the presence of truth as our visible friend and hourly companion' (in Furst 1980, 74). Associated with this was a series of dichotomies between science as rational, abstract, and detached; and poetry as dealing with emotions, feelings, and the values of the community. These contrasts informed the exchange between Peacock, the novelist, and P. B. Shelley. In 1820, in *The four ages of poetry*, Peacock provocatively suggested that poetry belonged to the infancy of civilization and must now give way to science. Shelley's *Defence of Poetry* of 1840 argued that the Poet was the moral voice of society against the encroachments of mechanistic thinking (Brett-Smith, 1947). Much later, Wordsworth had these oppositions in mind when accepting an honorary doctorate from Durham University: 'These things are not worth adverting to, but as signs that imaginative Literature notwithstanding the homage now paid to Science is not wholly without esteem' (Wordsworth to Henry Crabb Robinson, December 1838, in Morley 1927, I, 374).

Whewell's early discussions about the value of science took place within this Romantic framework. Robert Preyer has drawn attention to the network of friendships at Trinity that linked classical and German scholars such as Hare, Rose, Thirlwall, Maurice, and Sterling with men of science such as Peacock, Airy, Herschel, Sedgwick, and Whewell (Preyer 1981). Rose and Hare were particularly close to Whewell and discussed Romantic poetics and Coleridgean metaphysics with him from at least 1817. Since Preyer's emphasis is on the points of consensus among these people, it is important to say that Whewell's initial impression was far from favourable. He noted his reaction to Coleridge's *Biographia Literaria* in 1817 and to the author's 'peculiar sense of his peculiarity' (Todhunter 1876, I, 350-1; Diary 25 July 1817, WP, R 18. 9^2). In disputing Coleridge's metaphysical views with Rose he described them as 'muddy with their own turbulence'. The only good thing about the *Biographia*, he declared, was its criticism of Wordsworth's poetry (Whewell to Rose, 31 July 1817, WP, R 2. 99^6; Stair-Douglas 1881, 29). In a note on Coleridge's Lay Sermons he detected 'an attempt to set enthusiasm above reason', a theory designed to prepare reason for suicide:

The Reason of Coleridge seems so far as can be made out from the unintelligible mysticism of his language to be the principle of enthusiasm –

the principle by which things which cannot be proved are asserted, things which cannot be asserted are believed, and things float across the mind which can neither be proved, asserted nor believed. (Todhunter 1876, I, 349, 350-1)

Over the next three years this antagonism cooled and Whewell began to establish contact with the Romantic school. In 1821 he travelled to the Lake District with a letter of introduction to William Wordsworth from the poet's brother, Christopher, who had been appointed Master of Trinity in 1820. This had a positive effect and he told Rose that the poet was 'not half as Wordsworthian as his admirers'. He even referred to his 'lakification', and recommended works of imagination to his sister, Martha, saying that they were an admirable source of knowledge of 'the principles and feelings which appear in human nature' (Whewell to H. Wilkinson, 5 September 1821, and to his sister, 21 October 1821, in Stair-Douglas 1881, 66–7, 69). This contrasted with his recommendations in 1815 when he spoke of such 'standard works as Pope's "Homer", Milton's "Paradise Lost" (and above all Whewell's "Boadicea")!' (Whewell to his sister, 11 January 1815, in Stair-Douglas 1881, 15). Hare later felt confident that Whewell would agree when he wrote: 'It is indeed a delight to us both, that we shall hereafter associate those mountains and lakes, and our walks amongst them, with the most critical moments in our lives' (Hare to Whewell, 2 November [1841], WP, a. 77[139]).

The salient point here is not so much the extent of Whewell's agreement with the Romantics but the way they provided a set of oppositions that framed his early discussion of science. This dialogue was conducted on the assumption that Rose and Hare were members of the 'Wordsworthian' school, and that Whewell was not (Whewell to Rose, 22 September 1822 and 16 December 1823, in Stair-Douglas 1881, 93–4). In poetic theory, they saw Whewell as a defender of Pope, while Whewell was sceptical of the status of poetic criticism – Schlegel's, Coleridge's, or anyone's. He told Rose that when closely examined, 'most of the good criticism you see produces its effect rather as eloquence than as philosophy – rather excites poetical emotions than analyses them' (Whewell to Rose, 30 August 1817, in Stair-Douglas 1881, 31). This was significant because it seems to have been linked with a concern about the possibility of achieving certain conclusions in any sphere of knowledge. There was a sense in which his friends argued with Whewell as a practitioner of Cambridge mathematics and natural philosophy. This is discernible in Whewell's letter of 1819 where he contrasted Hare's belief that 'there is certainty to be obtained in matters of taste' with his own convictions:

You must allow that science is a much more satisfactory study; your knowledge there is undeniable and its accumulation eternal and imperishable. You know what truth is and you are sure that when you possess it no change of feelings can prevent your holding it fast and reaching it to future ages. (Whewell to Hare, 25 February 1819, WP, Add. MS. a. 215²)

This may well have been the confidence of a young man who had decided to compose a textbook on mechanics as an example of a successful science. But if this exercise was the one which, in Fisch's words, allowed Whewell to confront science as 'an object of study', it was also conditioned by his prior involvement in a debate in which science represented one alternative in a set of dichotomies.

Throughout this discussion Whewell did not want science to be constructed as the obvious antithesis to poetry, and hence to the realm of imagination, feelings, and emotions. He complained about the way in which Rose and Hare, following Coleridge, had set up a philosophical opposition between the rational and the affective. 'Why', he asked, 'will you not see that in speculative matters, though Reason may go wrong if not guided by our better affections, you cannot do without her?' The danger here was this: 'Finding that Reason alone cannot invent a satisfactory system of morals or politics, are you not quarrelling with her altogether, and adopting opinions *because* they are irrational?' (Whewell to Rose, 29 December 1823, in Stair-Douglas 1881, 95).

Another area of dispute was one in which science was identified with the search for utility and novelty. In a sermon of 1826 on *The tendencies of prevalent opinions about knowledge*, Rose charged that the current strength of pragmatic utilitarian doctrines was sustained by the popularity of 'Experimental Philosophy'; immediate results had become the object, and public opinion the arbiter of the value of knowledge. He also claimed that science placed undue stress on novelty for its own sake, regarding progress as 'the constant rejection of present belief in favour of new views' (Rose 1826, 12, 17). Whewell resisted these conceptions of science. First, he told Rose that he was puzzled by the equation of science and utility, indicating that 'all of the science which we learn here [at Cambridge] is perfectly devoid of all practical use'. Second, he argued that novelty was not in itself a criterion of scientific advance: 'the novelty, if the philosophy have been duly inductive, *includes* old truths and shews them from a new point of view' (Whewell to Rose 19 November 1826 and 12 December 1826, in Todhunter 1876, II, 75, 79). This dispute with Rose probably prompted Whewell to write his four sermons of 1827 which sought to show that science, and its intellectual processes, were compatible with religious views; these sermons formed the starting point of his Bridgewater treatise of 1833. But it is significant that these early considerations of science and its values derived from the need to answer the Romantic representation of science as opposed to poetry and the sphere of values. In the longer term, Whewell's response was to affirm the consideration of values within the philosophy of science, to show that science required a moral attitude (see chs. 5 and 7).

Whewell's encounter with Romanticism was not unusual. Indeed, William Rowan Hamilton, the Irish mathematician, seems to have worked out his commitment to science through a negotiation with Romantic ideas about poetry (Graves 1882–5, I, 193–4, 216–17, 271). Hamilton's case offers an instructive comparison with Whewell's. Trained in mathematics at Trinity College, Dublin, he spent his early years negotiating the twin attractions of poetry and science. He met Wordsworth in 1827 and corresponded with him frequently about the values and demands of these two vocations, a discourse he repeated with his aristocratic student, Viscount Adare, from February 1830. Like Humphry Davy, Hamilton felt suited to both these high callings, and began sending samples of his poetic efforts to Wordsworth for comment. The great bard saw promise in them but also offered stringent criticism. By 1829, with Wordsworth's help, Hamilton came to the conclusion that 'Poetry alike and Science are Muses that refuse to be successfully wooed by the same suitor' (Graves 1882–5, I, 315; Graves's wording). However, he continued to see them as the twin poles of his intellectual world: when he visited London in March 1832 it was to see first Coleridge and then Herschel (Graves 1882–5, I, 552, 538–42; Hill 1978–9, V, 535).

Whewell and Hamilton did not meet until April 1832 but the involvement of the younger mathematician with Romanticism offers a perspective on Whewell's earlier engagement. Unlike Whewell, Hamilton appears to have accepted the dichotomies constructed by the Romantics. In the mid 1820s he viewed poetry as a 'rival principle' which effectively balanced his passion for science; the former was congenial to the imagination and feeling, the latter required severe reason. Many women were eminent poets, but few were mathematicians (Hamilton to Miss Lawrence, 1825, in Graves 1882–5, I, 193; also 216). Wordsworth told him that the 'logical faculty' had much to do with poetry, but Hamilton continued to speak of them as related, but opposing activities. Wordsworth reinforced the drama of the choice by telling Hamilton that 'the renunciation of Poetry for Science' was a matter of great dignity (Wordsworth to Hamilton, 23 December 1829, in Hill 1978–9, v, 183; for Hamilton's epistemology, which also incorporated a dynamic dualism, see Fisch 1991a, 65; more generally, see Hankins 1980).

There is a further point of contrast. Whewell never saw himself as torn between the vocations of poet and scientist. His discussion with the Romantics, or at least the Wordsworthians, was not over such a choice, but rather over the image of science in their philosophy of knowledge. More so than Hamilton, then, Whewell realized that the contrasts between science and poetry in much Romantic writing demanded a philosophical and moral defence of the nature of science. It is possible that the interaction with Romanticism may have shown him how to be a 'critic'.

Whewell's engagement with Romanticism may have given him the opportunity of perceiving criticism as a distinct intellectual activity. In Romantic theory, the poet was both critic and creator (Eagleton 1984, 41–2). At first, Whewell was sceptical about the value of poetic criticism, writing a note to himself in September 1817 that 'Criticism can only spoil poetry,' and telling Rose that Coleridge 'has almost too metaphysical a head to be a good poet' (Todhunter 1876, I, 352; Whewell to Rose, 30 August 1817, in Stair-Douglas 1881, 31 and WP, Add. MS. R 2. 99[7]). At this stage he was reading both Pope and Schlegel but remained confused by rival principles of criticism. Later, in 1828, he told Wordsworth that 'a person may be too good a writer of verses to be a good critic of versification, or rather I should say a good anatomist of verse' (Whewell to Wordsworth, in Hill 1978–9, IV, 681). The question of the relation between criticism and practice was now confronting him – with respect to science and the contribution he could make. By this time, Hamilton had chosen his direction, acknowledging science as his calling. Whewell's situation was more difficult: he contemplated a vocation as a *critic* of science.

Apart from this connection with the English Romantics, Whewell was also familiar with the work of Friedrich Schiller, the German poet and critic. As Schaffer has recently suggested, the image of intellectual and social progress in Schiller's *Der Spaziergang* of 1795 may relate to Whewell's vision of the historical development of science. In 1842 he discussed Schiller's merits with Herschel, whose translation of this poem appeared in the volume of English hexameters Whewell edited in 1847 (Whewell 1847b; Schaffer 1991, 231). It is likely that Whewell read Schiller much earlier. He took notes on Schlegel in 1817 and from Madame de Staël's account of German literature, which included a chapter on Schiller, in 1821 (see WP, R 16. 14[18] on Schlegel and R 18. 9[8] on de Staël). I am not aware of evidence that would allow a more precise dating or acknowledgement of influence, but given the problem of available British models for the critical role Whewell was consciously seeking, at least by 1830, it is possible that Schiller may have been important.

In his *Letters on the aesthetic education of man* of 1794 Schiller diagnosed a split between the rational and affective faculties as the problem confronting civilized man. Like DeQuincey, he granted that the progress of knowledge required specialization: the 'increase of empirical knowledge, and more exact modes of thought, made sharper divisions between the sciences inevitable'. One consequence was that 'the intuitive and the speculative understanding now withdrew to take up positions in their respective fields'. But the fatal consequence was a lack of harmony in the character of individuals (Schiller 1967, 33–4). As suggested earlier, this dichotomy between reason and emotion, or imagination, was the subject of Whewell's dialogue with Hare and Rose. In conceiving this as a cultural problem, associated with scientific specialization, Schiller may have offered Whewell a vision of how a critic of science could legitimately present this study as one bearing on social and moral concerns.

A ROLE AS CRITIC?

When he wrote to Herschel in 1836 Whewell seemed to have clarified his vocation. 'In a year or two', he predicted, 'I expect to be a philosopher and nothing else' (Whewell to Herschel, 9 April 1836, in Todhunter 1876, II, 235). This was not a refusal to use the term 'scientist' which he had coined three years earlier; it was an acknowledgement that the role he had chosen was distinct from the scientific research in which Herschel was so heavily engaged, and even from his own academic duties and 'active business of the College'. Yet it was crucial for both science and undergraduate teaching: it was the 'reform of our Philosophy'. This, he declared, is the work 'to which I have the strongest vocation, and which I cannot not give up if I were to try' (1876, II, 234). Eight months later he spoke of his project for 'a history of all the physical sciences' as one that Herschel would appreciate because 'you have formed and executed a plan of the same kind'. Here Whewell acknowledged the encouragement of Herschel's *Discourse*, but added that: 'We shall not, however, have much ground in common. My scheme is so wide that all the life that is allotted me will be little enough for carrying it through' (Whewell to Herschel, 4 December 1836, in Todhunter 1876, II, 248–9). In the same period he confided to Hare and Rose his justification of this role as a moral one compatible with his position in an institution of the Established Church (see ch. 7).

Thus until 1837, most of Whewell's scientific contemporaries had only a partial account of his agenda as a metascientist. Herschel knew of it as a Baconian survey and analysis of the inductive sciences. Hare and Rose were aware of it as a concern with utilitarian morals and Romantic philosophy, and the association of both with the physical sciences. Probably only Jones, who began by discussing political economy with Whewell in the early 1820s, had a more complete sense of the ambitious project. Whewell outlined this to him in 1834, but as mentioned earlier, their discussions by this time had gone further.

Before the *History*, Whewell's public appearance as a scientific critic was as a reviewer and an author of reports for the British Association on the state of progress in different fields. Brewster appreciated the critical survey of mineralogy:

I know not how to thank you for the admirable, and correct account you have given of my labours in your Mineralogical report. There is no reward which can be compared with that of having his toil appreciated by those whose opinions must guide those of the age in which they live. (Brewster to Whewell, 10 June 1833, WP, Add. MS. a. 201[82]; also July 1825, no. 74, for his support of Whewell's bid to get the Chair of Mineralogy)

On a more general scale, and also after reading the report on mineralogy, Lyell said that he no longer regretted that Whewell had not become a 'giant' in 'one department' of science, or at least been content with 'two or three of the Arts and Sciences':

I have for some years come round to the belief that you have been exercising the calling for which Nature intended you, and for which she gave you strength and genius, and that you have given a greater impulse to the advancement of science among us by being a Universalist, and by mastering so much of Chemistry, Mineralogy, Astronomy, Geology and other branches, than you would have done if restricted to the perfecting of any one alone. (C. Lyell to Whewell, October 1836, WP, Add. MS. a. 208[124]; Todhunter 1876, I, 112, with punctuation not in original)

But significantly, Lyell then said that 'I know of no other position in which you could more effectually bring to bear on one point the diversity of your acquirements than by accepting the Presidentship [of the Geological Society]'. This is perhaps an indication that there was no *definite* role for a metascientific critic like Whewell – except in a voluntary and impermanent post.

Support such as this did not reassure Whewell about this role as a scientific commentator. He felt reluctant to accept the presidency of the Geological Society because of his lack of detailed knowledge and told Sedgwick that 'I am not the proper person for your generalissimo' (Whewell to Sedgwick, 6 October 1836, in Todhunter 1876, II, 246). But Sedgwick's advice is interesting:

Don't hesitate one instant, Mr. President elect. Did you not pick geological rubbish out of my eyes in 1820? Have you not figured in the mines of Cornwall? ... Have you not given the only philosophical view of ... [mineralogy] that exists in our language? Have you not written the best review of Lyell's system that has appeared in our language? etc. etc.? You are just the man we want. (Sedgwick to Whewell, 12 October 1836, in Clark and Hughes 1890, I, 464)

This litany suggests that Sedgwick regarded Whewell's early metascientific activities as a fair qualification for leadership of a scientific society. These were also listed by Lyell in his vote of thanks in 1838 for Whewell's term as president, although he referred to Whewell not as a man of science, but as an 'author and scientific workman' (C. Lyell to L. Horner 24 February 1838, in K. M. Lyell 1881, II, 38). As I have already hinted, Whewell aspired to something far more grand than this, but he did not make a full public statement of his claims until the *Philosophy* of 1840.

The next four chapters examine Whewell's pursuit of this metascientific role in different forums and through different genres. Chapter 4 considers the Victorian periodicals as one significant forum in which science was discussed before an educated readership larger than the membership of scientific societies. This was certainly not the only lay audience for science, but a focus on scientific reviewing in these 'high-brow' journals highlights the tensions in the public discourse on science. Whewell's first major interventions in scientific criticism took place in these journals where, as Sedgwick remarked, he had written 'the best review' of Lyell.

PART TWO

CHAPTER 4

Reviewing science

The being of an Englishman has no great cycle, which it would accomplish between the cradle and the grave: its longest revolution is performed between the quarterly publications of a review.

John Sterling 1848 [1828], II, 45

I SCIENCE AND REVIEWS

Walter Houghton, whose name is now rightly identified with the study of Victorian periodicals, has presented them as one of the most striking cultural phenomena of the century. The grounds of this assertion are threefold: the sheer number of publications (several hundred reviews, magazines, and weeklies), the constellation of prestigious contributors, and the kind of audience (Houghton 1982, 1). G. M. Young described the leading journals as written by and for the 'articulate classes, whose writing and conversation make opinion' (G. M. Young 1936, 6). The leading journals, beginning with the *Edinburgh Review* from 1802 and the *Quarterly Review* from 1809, constituted the intellectual forum of the nation, an aspect of the public sphere which Habermas has seen as characteristic of liberal bourgeois regimes from the eighteenth century. Historians of Victorian science are well aware of the presence of science in the major periodicals of the day, particularly by the time of the *Fortnightly* (1865) and *Contemporary* (1866) reviews, the crucial platforms for publicists such as Huxley, Spencer, Tyndall, and Lewes. But there is very little explicit analysis of scientific reviewing, even though it is recognized that men of science such as Playfair and Brewster wrote extensively in review journals, the latter depending on them for income. Before looking at Whewell's use of this medium, it is necessary to gain some sense of the review genre, the assumptions which supported it, and its significance as a location for metascientific discourse.

The important cultural function of periodicals was closely connected with their style and format. The most famous were primarily *review* journals, disseminating and evaluating an array of books, pamphlets, and addresses. Michael Wolff has suggested that they were the venue in which 'all the second speeches in the national debate were made' (Wolff 1959, 270). Their success was so great that some observers believed that during the nineteenth century the journal article attained priority over the book. Poole, the compiler of an index to Victorian periodicals, remarked in 1882 that the best writers now wrote articles rather than books or pamphlets (quoted in Houghton 1966–88, 1, xv). In 1846 the French historian Jules Michelet saw this as most pronounced in England: 'The English scarcely write anything now-a-days but articles in reviews' (Kent 1989, 1).

Since the inception of the *Edinburgh* and *Quarterly* reviews, the practice of considering several related (or even unrelated) books together had become common. This gave the talented reviewer the chance to impose his authority on the essay, to create the impression that the topic was his own construction and that the books listed at the head of the article were merely grist to his mill. Editors of the major quarterlies encouraged this promotion of their star performers. In his wonderfully satirical *Advice to a young reviewer* of 1807, Edward Copleston captured the illusion of authority created by clever reviewing, exposing its tricks and artifices, such as drawing selectively upon an author's preface to attack the book itself (Copleston 1807, 6–7). Until the latter half of the century, the anonymity of articles was largely unchallenged – until 1870 about 97 per cent were unsigned – and this encouraged intrigue about the author. The possibility of a famous person speaking on a controversial issue without the restraint imposed by their position was part of the attraction of this tradition. In many cases, of course, the identity of reviewers was known, but the notion of anonymity was also reinforced by the idea that articles represented the position of the journal, not the author, and this sustained an energetic rivalry between journals of different political allegiance such as the Whig *Edinburgh Review* and the Tory *Quarterly Review* (Maurer 1948; Shattock 1989, 15–17).

How did these features of the journals fit with the case of science? From the discussion in chapter 2, it is clear that the public space occupied by periodical literature was one that advocates of science could not afford to ignore. Thomas Young argued in 1810 that 'the *diffusion* of a respectable share of instruction in literature and the sciences' was more important to the public, than the discovery of new truths (T. Young 1810, 463). Ten years later the *Edinburgh Review* believed that this was under way, asserting that the extension of scientific knowledge to the public was more apparent in England than in France, where 'knowledge of every species is more confined to one class' (anon. 1820, 508). Although ahead of Britain in exact science, the upper social circles in France were seen as less in touch with science than their counterparts in England (Chenevix 1820, 389, 411). Later, in the controversy about the decline of science in England, the Dutch professor Gerrit Moll praised British science for being less specialized and more accessible than its French counterpart (Moll 1831, 14). The success of scientific leaders in using the periodical press was acknowledged even by those who were suspicious of the cultural pretensions of the scientific enterprise. Thus in *The idea of a university* (1852), J. H. Newman accepted the value of 'that superficial acquaintance with chemistry, and geology, and astronomy, and political economy ... which periodical literature ... diffuse[s] through the community' (cited in Houghton 1982, 8).

The charter of the British Association incorporated this recognition of the wider audience of science, even though it simultaneously enabled the consolidation of experts (Yeo 1981, 72–8). Its critics were able to exploit the assumption that scientific ideas should be accessible. Thus Dean William Cockburn, Bishop of York and scourge of clerical geologists such as Buckland and Sedgwick, protested in these terms:

I have the right to complain of philosophers who put forth a proposition of general interest in such terms that a man of common understanding cannot tell what they mean. I have a right to call upon the assembly at Newcastle to publish some more distinct explanation of this new and obscure discovery. (W. Cockburn 1838, 11)

Baden Powell believed that men of science must accept an educational role; and in his account of the undulatory theory of light, to which Cockburn was referring, acknowledged that there was a 'constant necessity for a vigilant review and connected recapitulation of ... [theories] from time to time' (Powell 1841, 1; Corsi, 1988, 163). But although the journals were successfully used to disseminate science, there were some particular problems associated with science reviews.

AVAILABLE BOOKS

In the case of major contributors such as Jeffrey, Macaulay, and Mill, the review essay was a vehicle for their own preoccupations. All they needed to do was find a book, or a number of them, on which to hang the article. In areas such as theology and literature this was not usually a problem. But since many authors were writing in periodicals precisely because they could not get books published – Carlyle is the notable example – a lack of appropriate books may not have been uncommon. In the case of political economy during the 1820s, J. R. McCulloch, one of the leading writers on this subject in the *Edinburgh Review*, was compelled to create fictitious titles in order to maintain the appearance that his articles were reviews (Houghton 1982, 24). Brewster admitted that he had promised to write an article on the decline of science for the *Quarterly Review*, 'but there was no book till Babbage's appeared, which could authorize the insertion of such an article' (Brewster to Forbes 10 July 1830, in Morrell and Thackray 1984, 27).

There is reason to expect that the scope of science reviewing may have been restricted by the lack of suitable books. In explaining the need for specialist scientific journals, the *Annals of Philosophy* in 1813 acknowledged the emergence of the 'Monthly and Critical Reviews', but noted that since they were 'entirely confined to criticisms of *books*, they could scarcely be considered as registers of the discoveries in science' (anon. 1813, 2). That is, books on science were scarce, and most reports on experiments and discoveries were published in the transactions of scientific societies. Lyell remarked on the 'limited sale of scientific works, even of profound research', reporting that publishers were reluctant to take them 'at their own risk, even when proceeding from authors of acknowledged talents' (C. Lyell 1826, 167). In 1842, Brewster referred to the difficulty of publishing books on abstruse physical science and explained that this was why there were so many original articles on science in encyclopaedias (Brewster 1842b, 53). Even a successful publisher such as John Murray was sceptical about the market for books on science. When Brougham suggested a print run of 1,500 for Mary Somerville's *Mechanism of the Heavens*, Murray printed 750, only to find that these sold out, but still welcomed Somerville's intention to make the next book 'more popular' (Murray to William Somerville, 2 August 1830, in Smiles 1891, II, 406–7; also 408. Mary Somerville's husband handled the financial arrangements).

It is difficult to assess the representation of science in the major quarterly reviews. Compared with politics and literature, its profile was at best moderate, and its reviewers could not attract the following of the renowned contributors, such as Macaulay, who wrote forty articles for the *Edinburgh Review* between 1824 and 1844. Brock reports that in 1824 out of 208 articles in the *Quarterly* and *Edinburgh* reviews, and *Blackwood's Magazine*, only two explicitly concerned scientific matters; but, as he observes, this was an unusually low proportion (Brock 1988, 48). Between 1804 and 1819, John Playfair wrote at least fifty articles in the *Edinburgh Review*, most of which dealt with mathematical and physical sciences, others with geology, palaeontology, or travel books that raised scientific issues. As Morrell suggests, this assertion of the 'importance of science in general literate culture', has been missed by most historians of the quarterlies (Morrell 1971a, 56).

By the second quarter of the century the demand for accounts of science in these journals was substantial enough to allow Brewster to rely on reviewing for a large part of his income, writing twenty-nine articles between 1833 and 1844 in the *Edinburgh Review* and seventy-seven between 1844 and 1863 in the *North British Review*. This effort was supplemented by contributions to the *Quarterly Review*, *The Foreign Quarterly Review*, and the *Monthly Chronicle*. Brougham easily rivalled Brewster, penning some 320 articles for the *Edinburgh Review* between 1802 and 1854. In Brougham's case of course, science was not the only subject covered, although scientific matter featured in many of his articles on education. The contributions of Whewell, Herschel, Lyell, Powell, and De Morgan – to name some of the most significant science reviewers – were far less numerous, but the collective effect was substantial.

How significant was this in statistical terms? This measurement depends partly on what is counted as a scientifically based article. Subjects such as travel literature, medicine, mechanical arts, are borderline cases here, for although they did involve scientific issues and were covered by authors such as Playfair and Brewster, they belonged to different cultural niches than say astronomy, chemistry, geology, or physiology. The leaders of the British Association were well aware that these latter subjects were in greater need of careful

Table 1. *'Science' articles as a percentage of total articles between 1830 and 1840*

	Total number of articles	'Science' articles
Edinburgh Review	473	45 (9.5%)
Quarterly Review	450	53 (11.7%)
Westminster Review	576	25 (4.4%)

public promotion. Nevertheless, it is possible to form some impression by checking the proportion of 'scientific' articles in the three leading periodicals between 1830 and 1840 (see table 1).

This is not a high proportion compared with subjects such as politics and literature, but it is significant when the lower number of scientific books is considered. It has been estimated that of roughly 45,000 books published in England between 1816 and 1851 the largest category was easily theology/religion (10,000), the next largest history/geography (4,900), ahead of fiction (3,500) (R. M. Young 1985, 134–5; Cruse 1930).[1] Not surprisingly, then, books on science were a sectional minority. Given this, the proportion of science-related reviews in the main quarterlies may be higher than expected. If so, it is possible to view them as a crucial forum through which science, a culturally marginal activity, sought to increase its status. As we shall see, this is certainly a viable interpretation of what writers such as Whewell and Brewster, in different ways, were doing in their contributions.

ANONYMITY

Although scientific topics were reasonably well represented in the journals, there are reasons to expect tensions between the genre of the review article and the particular case of science. First, as noted above, the convention of anonymity played on the possible identities of authors speaking on controversial topics. But this depended on the existence of well-established debates with known positions, such as those on the Corn Laws, Catholic emancipation, or the admission of

---

[1] Kelly (1966), 519, 524 gives the proportion of science books in lending and reference libraries after 1876 as *c.* 11 per cent, even when grouped with art and education. This compares with 23.5 per cent for history/geography/biography; 17 per cent for literature; by this date theology represented only 5.4 per cent.

Dissenters to university. Could this be carried over into science in the early nineteenth century when scientific institutions aimed at ensuring that their pursuits were recognized as non-threatening cultural practices? Second, anonymity was supported by the belief that articles represented the position of the journal. But how could this apply to science unless it became linked with political and social issues, thereby risking controversial involvement? Journals rarely held an editorial position on a substantive scientific question – although Brougham's use of the *Edinburgh Review* to denounce Young's optical theories must be noted – but both the *Edinburgh* and the *Westminster* reviews attacked the Cambridge and Oxford curricula for their lack of science. Cannon warned against simplistic predictions here, suggesting that the politically conservative *Quarterly Review* was more 'liberal' on scientific matters than the Whig *Edinburgh Review* (Cannon 1978, 229). It is true that Brewster wrote on the 'Decline of science' in the *Quarterly Review*, and Lyell used it to promote scientific institutions and university reform. But on the other hand, Brewster also attached science to a reform programme in the *Edinburgh Review*, a concept Whewell carefully avoided, or resisted, in writing for the *Quarterly Review*.

PROBLEMS OF STYLE

Reviews were meant to entertain, and the best writers did so in 'trademark' styles. In the case of science, however, there were some special problems with the balance between technical exposition, criticism, and readability. Some topics were difficult to enliven; others were too sensational. For example, Richard Owen sought advice from Murray on the inclusion of a diagram illustrating the 'reproductive economy and apparatus' of the bee, in an article on parthenogenesis (cited in E. Richards 1989, 258). But the more chronic problem, in the opinion of editors, was the shortage of willing and competent reviewers. John Lockhart, the editor of the *Quarterly Review* from 1825 to 1853, and his publisher, Murray, were interested in science, geography, and expeditions. They were both friends of Lyell and Murchison, and wanted scientific topics covered (Smiles 1891, II, 390–3). But Lockhart complained to Murray about the difficulty of finding a 'scientific hand':

Oh! if we had but a first-rate man of business, who could and would write clearly and briefly – a Playfair or a Davy, or a Brougham, with all his blunders and superficialities! Any hand that could command attention and give pleasure with instruction, however imperfect. Our Whewells, Brewsters, Lyells, etc., are all heavy, clumsy performers; all mere professors, hot about little detached controversies, but incapable of carrying the world with them in large comprehensive *resumés* of the actual progress achieved by the combined efforts of themselves and all their rivals. (Lockhart to Murray, 26 October 1839, in Smiles 1891, ii, 454)[2]

This suggests that it was difficult to translate science into the acceptable review style. Lockhart implied that a good general writer reviewing science might be preferable to a specialist scientific reviewer. However, there were dangers here: as Copleston warned in 1807, the standard reviewing tricks could fail if the reviewer tried to 'meddle with works of deep research and original speculation ... [which] cannot be treated superficially without fear of being found out' (Copleston 1807, 7). This may explain why scientific reviewing demanded the special abilities Lockhart found so scarce, since the writer had to understand the topic and then relate it to a non-scientific audience in an interesting manner. Thus in terms that could hardly be described as eloquent he wrote to Whewell in 1834: 'I really am in a scrape unless the next Quarterly shall have some notice of Mrs. Somerville's new book, and I know not, Herschel being away, and Brewster engaged for the thing in the Edinburgh, to whom to turn unless to yourself' (Lockhart to Whewell, 24 January 1834, WP, 266 c. 80 149[24a]).

Within the scientific area, Lockhart was content to rely on authors he knew rather than to seek those with particular competencies. Thus when thinking of an issue devoted to biography he mentioned Herschel as the best person to write about Cuvier (Lockhart to Milman, 1 July 1831, in Lang 1897, ii, 99). But even with an author of Whewell's abilities, the appropriate mix of tone and content was a delicate matter. Lockhart was not convinced of the 'probable popularity' of Whewell's review of Herschel's *Discourse*, and asked Murray's comment on the reaction of 'the reader in general'. He then advised Whewell to use more quotations of an 'interesting character', suggesting that by making it 'clearer and more entertaining, you will of course extend the circle of its readers' (Lockhart to Whewell, 1 March 1831, WP, O 15. 46[11]). Lyell admitted that it was

---

[2] Whewell surmised that 'we *freshmen* reviewers are too serious for Lockhart: if I ever review again, I think I shall know my trade better than before' (Whewell to Murchison, 28 April 1833, in Todhunter 1876, ii, 164).

more difficult to write for 'general readers than for the scientific world', and was annoyed with those who 'think that to write *popularly* would be a condescension to which they might bend if they would' (C. Lyell to Dr Fleming, 3 February 1830, in K. M. Lyell 1881 I, 260; Porter 1982, 34). There was a judgement that the readers of reviews wanted scientific topics related to wider philosophical or religious issues. This may also have been the view of some men of science, since Buckland complained about the treatment of his *Vindiciae Geologicae* (1820), saying that he expected 'something higher ... and more philosophical' than the technical and scientific manner in which it had been done by John Barrow. The balance here was not easy to strike (Shine and Shine 1949, 84).[3]

ATTITUDE OF MEN OF SCIENCE TO REVIEWS

In spite of these tensions between editors and contributors, the leading scientific figures of the early nineteenth century recognized the importance of the periodical journals. This took various forms. As we shall see, most acknowledged the value of a favourable review, not only as a means of extending their work to a wide audience, but also as a public adjudication by one of their peers. Herschel, Jones, Lyell, and Somerville all had this attitude towards Whewell's essays on their books. At another level, Lyell, according to Samuel Smiles, believed that writing articles for the *Quarterly Review* 'helped to prepare his mind' for work on the *Principles of Geology*, 'without exhausting his materials' (Smiles 1891, II, 267). Mark Pattison later remarked that 'books now are largely made up of republished review articles' (Houghton 1982, 21). But as well as promoting or facilitating scientific ideas, reviews were also seen as a space in which different things could be said. Lyell instructed Poulett Scope, about to review the *Principles*, to point out 'the moral' about Moses more clearly than he had done: 'It is just the time to strike, so rejoice that, sinner as you are, the Q.R. is open to you' (C. Lyell to Scrope, 14 June 1830, in K. M. Lyell 1881, I, 270–1, also 300; Porter, 1982, 39). But this enthusiasm for anonymous articles as a vehicle of anti-clerical polemic was not widely shared among men of science.

---

[3] Brewster judged that 'the taste of the day is bent upon popular and exciting articles' which were not easy to procure, but had to be sought, because 'no scientific writer need expect the public to take an interest in profound Science' (Brewster to Napier, 5 February 1833, BL, Napier MSS. 34, 616, f. 21).

What was generally agreed was the desirability of using the space they provided as a means of improving public understanding of science and the status of scientists. Writers such as Whewell, Brewster, Powell, Herschel, and De Morgan all regarded this as an urgent need. They would have been shocked by the opinion of a correspondent of Murray's in 1824 who felt that the advance of science was uninterrupted, and periodicals were merely needed to supply 'the public with the facts' (Francis Cohen to Murray, 26 August 1824, in Smiles 1891, II, 161–2). When Brewster reviewed Herschel's *Treatise on Sound* in 1831 he lamented the way in which science had been 'excluded from the accomplishments which qualify for public life', and suggested that the level of scientific knowledge among the 'educated classes' had to be increased. Then 'our reviews, magazines, and journals would be induced to devote a portion of their pages to the development and simplification of modern discoveries' (Brewster 1831a, 475–7). Throughout his entire career, Brewster cited examples of the failure of science to raise public understanding: fear of comets, phrenology, almanacs, anti-Newtonianism, and *Vestiges*, etc. In 1838 he reported on the manner in which London had been seized by a 'water panic': rumours of pollution had spawned filtering machines and other gadgets. 'It was in vain', he concluded, 'that Brande analysed the water at the Royal Institution, and Faraday attempted to lecture London into its senses. Knowledge ceased to be power; philosophy lost its authority' (Brewster 1838b, 76).

This issue of intellectual authority was recognized by the major advocates of science, and most agreed that all opportunities to affirm it had to be seized. Indeed some writers such as Babbage, Powell, Airy, and De Morgan did not restrict their activities to major periodicals but wrote for popular encyclopaedias and the series sponsored by the Society for the Diffusion of Useful Knowledge, and gave lectures to various audiences. Both De Morgan and Powell contributed anonymous articles to the *Quarterly Journal of Education*, a short-lived entity sponsored by the Society for the Diffusion of Useful Knowledge, in which they advocated science as part of university education and attacked the conservative attitudes of Cambridge and Oxford. Powell also edited and wrote for the *Magazine of Popular Science* in 1837 (De Morgan 1832b; Powell 1834b; Corsi 1988, 5–6, 311). Several well-known scientists contributed to Dionysius Lardner's *Cabinet Cyclopaedia*, although Lardner told Herschel that his aim of producing scientific articles more 'general and popular' than those in the *Encyclopaedia Metropolitana* was thwarted by the trouble of attracting suitable scientific authors: 'My chief difficulty in Science is to find profound men who like yourself are able and willing to write a popular work' (28 July 1828, HP, vol. 11, no. 108). There were, however, some doubts about the tension between the advancement of science and its diffusion. Thomas Galloway wondered whether Herschel would not be better employed in further astronomical research than in producing popular texts (Galloway 1833–4, 165; also Yeo 1981, 1984).

2 WHEWELL'S EARLY REVIEWS

The second, and largest, part of this chapter deals with Whewell's early metascientific commentaries. These were made in review articles between 1831 and 1834 and therefore either preceded or paralleled his addresses and reports to the British Association – interventions which are now recognized as significant in the establishment of his position as an authoritative scientific critic. In 1831 he wrote four major reviews in leading periodicals: on the state of science in English universities, on the first volume of Lyell's *Principles of Geology*, and on Jones's *Essay on the Distribution of Wealth* – all in the *British Critic* – and on Herschel's *Preliminary Discourse on Natural Philosophy* for the *Quarterly Review*. He also reviewed the second of Lyell's volumes in 1832, and Somerville's *Connexion of the Sciences* in 1834, both in the *Quarterly Review*. In these, he was able to take advantage of the general and comparative criticism encouraged by the review genre. Endorsing this practice, he told Napier, the editor of the *Edinburgh Review*, that he did not object to a review that 'made the title of my book [on university education] the occasion of publishing an Essay on a subject only slightly connected with mine' (Whewell to Napier 1836, in Whewell 1838a, 187). This was the licence that Whewell was able to exploit in his own essays on other works. Whereas Copleston, in his satirical advice, warned that a reviewer must always follow 'public taste', Whewell sought to direct it (Copleston 1807, 3). This was the role he seized in 1831, aided by topics such as the place of science in universities, the progress of two new sciences, geology and political economy, and Herschel's treatise on the aims and method of natural philosophy. These topics easily supported general reflections on the nature of science, and Whewell's reviews constituted a forceful presentation of himself as one of the leading scientific writers of the day. Before looking at these articles, it is worth considering his attitude to some of the issues raised in the preceding discussion of the periodical press and its connection with science.

As early as 1817 Whewell and Rose had discussed the idea of founding and editing a journal to carry their views and those of their young friends, such as Hare, Herschel, and Jones. But Whewell had the sense to see that a group of young men might not have the breadth of knowledge to cover the topics included in the best reviews of the day: 'there are so few subjects', he confessed to Rose, 'on which I am not on the verge of ignorance that it would be of much more use ... that I should try to learn than to teach'. He was also critical of the effects of periodicals, telling Rose that 'people read reviews at present to spare themselves the trouble of reading original books and forming their own opinions' (Whewell to Rose, 14 September [1817], in Todhunter 1876, II, 17, also 18–21; Whewell to Jones, 16 October 1817, in Todhunter, II, 22–4).

In addition to these worries, Whewell, like Galloway, had doubts about the benefits, or at least the means, of popularizing science. He certainly resisted Lardner's efforts to capture him as a contributor to the *Cyclopaedia* series, and he had nothing to do with the other publications which gave some of his colleagues, such as Powell, Herschel, and De Morgan, access to wider audiences (Lardner to Whewell, 31 May 1831, WP, Add. MS. a. 208[4]). He told Jones that Herschel would do something important in science 'if he will give up spinning his entrails out into Encyclopaedias for such fellows as Lardner' (Whewell to Jones [November 1830], WP, Add. MS. c. 51[91]). Whewell restricted himself to books and major journals; he did not give general lectures until later in his career, and admitted in a lecture at the Royal Institution in 1854 that 'being so infrequently in this metropolis, I do not know what trains of thought are passing in the minds of the greater part of my audience, who live in the midst of a stimulation produced by the lively interchange of opinion and discussion' (Whewell 1854a, 3). It is also important to stress that his contribution to some of the leading journals during the 1830s was his first foray into an area with an audience extending beyond those who read university textbooks and sermons. This changed in 1833 with the publication of his Bridgewater treatise on natural theology. But even when he was in a campaigning mood, as he could be when talking about inductive philosophy with Jones, Whewell had a fairly selective notion of 'the people': as Simon Schaffer has noted, the readers of the *Quarterly Review* were not the people who were attending lectures on phrenology and other subjects in Mechanics Institutes (Schaffer 1991, 213–14).

Thus in looking at Whewell's early reviews, we need to realize that he was aware of at least some of the conditions associated with this important contemporary form. As well as analysing their content, this section of the chapter will also try to consider the problems of discussing science in this medium.

Whewell did not invent this mode of raising issues about the nature of science or promoting the character of the scientific enterprise and defining it for a general readership. Other men of science had exploited the review essay in this way: for example, Playfair on Laplace in 1808 and 1810, and on Cuvier in 1811; Lyell on scientific institutions and English universities in 1826–7; Brougham on science and popular education in 1826, Powell in the *British Critic* from 1823 to 1825; Brewster on the decline of science and the British Association in 1831 and 1833. These were all significant interventions. However, Whewell's concentrated period of reviewing in the 1830s was a distinctive contribution to this medium, perhaps the most influential mode of discussing the nature and image of science. It is appropriate therefore that some of the most significant examples of this form were in fact accounts by others of his own books, such as the astonishing *tour de force* by Herschel on the *History* and the *Philosophy* in the *Quarterly Review* of 1841, and Brewster's notorious articles on these two works in the *Edinburgh Review* in 1837 and 1840.

Whewell's essays involved a treatment of the specific works under review but, in the style of famous literary and political reviewers such as Jeffrey, Brougham, and Macaulay, they also operated at a more reflective level. This was sustained, in Whewell's case, by several recurring themes concerning the meaning of 'science' both within the scientific community and for the public. Addressing these two audiences was a significant and challenging task and, writing to Herschel, Whewell said that he considered himself 'one of the most fortunate men of the age' in being able to comment on both Herschel's *Discourse* and Jones's work on rent in the same year (Whewell to Herschel, 10 February 1831, in Todhunter 1876, II, 115). When the review of Lyell's *Principles* is added we can appreciate the responsibility which Whewell felt, for this was the moment in which a general treatise on the aims and methods of the scientific endeavour, and theoretical works on two of its most recent examples – political economy and geology – were brought to the attention of the educated reading public. All three reviews appeared in conservative journals: the Tory *Quarterly Review* and the High Church *British Critic*. Whewell had a link with the latter through his friendship with H. J. Rose, who was also one of the editors of another Anglican journal, the *British Magazine*. As indicated earlier, there was nothing unusual about reviews of science in these works. However, it did mean that they would be read by a conservative audience as the judgements of an influential Cambridge don on the intellectual and cultural standing of modern science.

An awareness of this audience probably conditioned one of the articles published by Whewell in January 1831. Ostensibly a review of the transactions of the Cambridge Philosophical Society, it appeared in the *British Critic* under the title 'Science and the English universities'. The attacks in the *Edinburgh Review* on the ancient English universities, which included an allegation of their neglect of 'modern knowledge', were part of the background (Smith 1810; Hamilton 1831a and 1831b). Speaking especially for Oxford, Copleston had replied to the 'three giants of the North' – as J. H. Newman dubbed Playfair, Smith, and Jeffrey (Copleston 1810; Mansbridge 1923, 152). By the time Whewell was writing, there was an additional assault from the *Westminster Review* (see anon. 1828).

In responding, Whewell explained that the 'more distinguished members of our Universities' had not replied to these accusations, feeling that such public brawling was undignified. He regarded this as an unwarranted luxury in 'stirring and angry times', and declared that the place of the universities in the 'scientific character' of the country should be more widely known (Whewell 1831b, 71). One effective answer to the charge that these institutions were outdated, he suggested, was the observation that two of the *'newest'* sciences – geology and the recent wave theory of light – were keenly pursued by members of both Cambridge and Oxford. In the case of geology, he declared, these were the only places in England where a student could learn from modern practitioners of the subject (1831b, 74). By the time he came to mention the work in physical science carried out by members of the Cambridge Philosophical Society, such as Airy, Lubbock, and Willis, Whewell felt that he had rebutted the 'unfounded attacks of a perverse generation of critics' (1831b, 90).

WHEWELL VERSUS BREWSTER

There was another stimulus to Whewell's formulation of this topic: namely, the recent debate stimulated by Charles Babbage's *Decline of Science in England* of 1830. Whewell did not refer to this work but it is likely that Babbage's lament about the absence of a 'profession' of science in England caused him to clarify the place of scientific research in the activities of a Cambridge Fellow. In his review of Babbage in 1830, Brewster said that in the eight universities throughout Britain no one was 'engaged in any train of original research' (Brewster 1830, 327). Whewell took exception to this in a letter to Brewster, and the article was a public clarification of the fact that the primary function of the universities was academic instruction, not the discovery of '*new*' truths. Nevertheless, he observed, some professors managed not only to 'communicate the existing stores of knowledge' but to pursue 'original' research (Whewell 1831b, 74–5). This did not satisfy Brewster, because for him the notion of a '*train of original research*' had a social meaning; it meant full-time pursuit of inquiries at the boundaries of knowledge and was thus incompatible with the varied duties of a Cambridge professor (see Brewster to Whewell, 4 November 1830, WP, Add. MS. a. 201[79]; Morrell and Thackray 1984, 29–30). He suggested that there be a definition of two roles: a lecturing professor and a philosopher engaged in research (Brewster 1830, 328). Whewell scoffed at the utopian financial assumptions of this scheme, but he was also suspicious of its implications for the idea of a university centred on undergraduate mental and moral training. In any case, an issue of the *British Critic* was not the place to consider the possibility that an educational institution identified with the Church could give *primary* attention to the production of novel research, especially since at this time originality was not deemed an appropriate element in the education of young men (Rothblatt 1985, 69). As recently as 1826, Whewell had been required to defend what Rose saw as the undue emphasis on novelty in experimental inquiry. In 1834 Powell pleaded for more space for physical science at Oxford, but he did this in a journal backed by the utilitarians. Bearing Whewell's audience in mind, we can say that he went as far as he could to affirm the presence and value of science within Cambridge.[4]

[4] Whewell's statement of 9 December 1828 to members of the Senate recommended more space for scientific collections in museums at Cambridge (WP, 266 c. 80[49]; and Becher 1986).

The comparison of Whewell and Brewster is quite useful in revealing different styles of reviewing. The political platform of journals was not the only factor: in this case the difference had more to do with the author and the image of science he promoted. Whewell was as much restricted in what he could say on scientific issues by his situation in Cambridge as by his choice of journals. Thus, even if he wished to, he could not have promoted science by stressing its practical, technological, or commercial value, since this would have raised concerns about the nature of liberal education (see ch. 8 below). Brewster, on the other hand, was a free-lance man of science who wrote partly for financial reasons, but also to proselytize his idea of science as an agent in a progressive social and cultural programme. He said much the same thing in any journal, telling the readers of the *Quarterly Review*, for example, that Britain was a third-rate scientific country in terms of institutions (Brewster 1828, 15), and warning in the *Monthly Chronicle* that scientists must explain science to the public if it was ever to become part of general culture (Brewster 1839, 113; also 1858, 179). But he was also able to link this campaign with that conducted in the *Edinburgh* and *Westminster* reviews against the ancient universities. This was what Whewell was responding to in the *British Critic*.

REVIEWING INDUCTIVE PHILOSOPHY

The reviews of Herschel, Lyell, and Jones were Whewell's first chance to put his views on science in a public forum beyond the scientific and university communities. Each work gave him the chance to address a lay audience on the intellectual and cultural value of the sciences, while keeping science divorced from the kind of social and political issues raised by Brewster. At the same time he was also able to assert his views on the method and philosophy of science against those of other commentators.

In the article on Herschel's *Discourse* of July 1831, Whewell believed that he was attempting nothing less than an account of the meaning of science and, for him, its keyword – induction. In the two other reviews he appraised the scientific credentials of geology and political economy. As we shall see, there was a latent tension here between the general idea of science and its particular exemplification in different disciplines.

Herschel's 'Discourse'

Whewell's review of the *Discourse* was difficult to write and he admitted his dissatisfaction with the result. One reason for this was that Herschel's book represented a novel genre for the English reader – a supposedly popular text on 'the history and philosophy of physical science'. In contrast with the Continent, where men such as Cuvier and Berzelius surveyed the progress of science, no individuals of similar authority had sought to extend the results of science to 'a wider circle' or to connect them with 'the general body of knowledge'. But Whewell noted that the *Discourse* contained not only a survey of the state of natural knowledge but an account of the principles on which it rested and the 'maxims by which its researches have been and must be successfully conducted'. Furthermore, it attempted this in a new way. While 'volumes upon volumes' had been written on 'the nature of human knowledge and the laws of human thought', there had been inadequate attention to the mental processes exhibited in the progress of science. Herschel's *Discourse* was thus 'one of the first considerable attempts to expound in any detail' the rules and doctrines of successful scientific method (Whewell 1831a, 374–7). As such, it was a welcome alternative to the treatments of induction in general treatises on logic, such as those of Whately.

There was a polemical aspect to this contrast, since Whewell, together with Jones, was annoyed by Whately's views on political economy and his insistence on embracing induction within syllogistic logic (Jones to Whewell, 24 February 1831, WP, Add. MS. c. 52[20]; also Corsi 1987). But, more generally, Whewell was referring to the fact that issues of scientific method had not been given separate and detailed treatment; rather, they were incorporated in more general treatises on metaphysics and epistemology, such as those of the Scottish philosophers. He was confronted with the task of composing a popular essay for a general audience on a subject which he saw as a new field, rather than one in which a consensus had been established.

It is possible to classify Whewell's review into points of agreement, difference, and silence. Not unexpectedly, the overall tone was positive: Whewell praised the *Discourse* as the remedy for what he regarded as an appalling confusion about the meaning of Baconian method and some of its crucial terms, such as 'induction'. Scottish writers such as Reid had contributed to this by speaking of 'the *inductive principle'*, by which they meant the 'instinctive belief in the permanent uniformity' of the laws of nature. Others envisaged induction as simple collection of data. Herschel, however, offered an extended account of induction as 'the process of considering a class, or two associated classes of phenomena, as represented by a general *law*, or single conception of the mind'. Second, by providing concrete examples of this method of investigation, Herschel was able to show the 'mutual dependence and contrast of induction and deduction' in the process by which the 'vast structure of science' was built up. Through these careful examples Whewell hoped that Herschel had made it less likely that the public would confuse purely deductive sciences – or those 'consisting in the consequences of a few axioms' and governed by syllogistic logic – with those that rightly claimed the 'name of inductive sciences'. Third, Herschel's discussion gave due emphasis to the importance of mathematics in allowing the expression of precise quantitative laws. Fourth, it *partially* recognized the way in which a 'distinct terminology' formed the language of science and enabled the transmission of the labours of one generation to the next (Whewell 1831a, 380–91).

Throughout the first two sections of the *Discourse*, Herschel sought to analyse and codify the method by which natural science had advanced: 'What we have all along most earnestly desired to impress on the student is, that natural philosophy is essentially united in all its departments, through all which one spirit reigns and one method of enquiry applies' (Herschel 1830, 219). This affirmation of an identifiable method as the clue to the success of science arguably avoided contentious epistemological questions and, judging from Whewell's response, Herschel's account of the salient features of inductive method provided at least a general point of agreement. Beyond this, however, there were tensions between Herschel and his friendly reviewer.

Whewell's position can be more easily understood if Herschel's approach is outlined. In his third section, Herschel acknowledged that while 'one method of enquiry' pervaded all science, it was necessary to look at 'the subdivision of physics into distinct branches, and their mutual relations' (Herschel 1830, 221). The chapters in this section of the *Discourse* are primarily organized in terms of objects of study in nature, moving from forces and the structure of objects to the motion of sound and light, then to cosmical phenomena, back to the constituents of the earth, and concluding with imponderable forms of matter. This produced a discussion of the sciences concerned with these natural phenomena, beginning with pneumatics and hydrostatics, moving through optics, astronomy, and geology and ending with heat, magnetism, and electricity. But while it is clear that Herschel regarded Newtonian astronomy, with its incorporation of the demonstrable laws of dynamics, as the most perfect of the sciences, the concept of a hierarchy was not obvious in his survey of the branches of science. When this image did appear it came in an unexpected form. Thus Herschel remarked that 'geology, in the magnitude and sublimity of the objects of which it treats, undoubtedly ranks, in the scale of the sciences, next to astronomy' (Herschel 1830, 272, 287).

Rather than drawing strong distinctions between subjects, Herschel was more concerned with showing how inquiries in different fields could overlap to produce more general laws. He hailed the convergence of magnetism and electricity, 'which had long maintained a distinct existence', as a great event, and suggested future useful interactions between botany and chemistry and botany and geology (Herschel 1830, 324, 345). Herschel stressed that 'no natural phenomenon can be adequately studied *in itself alone*, but, to be understood, must be considered *as it stands connected with all nature*'; and he expected that this idea of unity would be realized in 'higher orders' of generalization embracing several sciences (1830, 174, 259, 360).

In contrast, Whewell's account presented an image of science which stressed difference rather than unity or interaction. He thought readers would be interested to see the sciences arranged in their order of growth from those just beginning to form inductive generalizations to those which had reached what Herschel called 'fundamental laws' or, more figuratively, 'from the tottering girl to the full-formed matron' (Whewell 1831a, 390). Although Whewell explained that limits of space prevented a detailed comparison with Herschel's survey, some interesting contrasts are nevertheless apparent. Thus the 'nursery and spelling book' stage was occupied by botany, chemistry, and mineralogy; these were only just past the 'outset of their inductive career'. The sciences under the heading 'physics' – for example, magnetism and electricity – were further advanced because portions of their findings had been reduced to 'mathematical formulae'. Optics, Whewell's favourite 'modern science', had made rapid progress and was close to the top of the hierarchy. At the apex, of course, were mechanics and astronomy, representing the only instances in which the combination of induction and deduction had been successfully completed; in the case of astronomy, there was no chance that its central theory could be 'overturned' (1831a, 390–7).

Now the criterion behind this classification – scope and quantification of laws – was certainly present in the *Discourse*, but Herschel did not use it to draw the same kinds of distinctions of status between the sciences. Thus while Whewell relegated chemistry and mineralogy to the 'infancy' of the inductive project, Herschel saw them as fundamental, rather than low, in that they dealt with the basic 'material constituents of the world' (Herschel 1830, 290). Both these subjects concerned the most basic natural 'substances, a knowledge of which necessarily preceded many other sciences' (1830, 143, 290–9). While Whewell omitted geology from his hierarchy, explaining that it raised additional questions, Herschel gave it high praise, even illustrating the meaning of *verae causae* with examples from this 'deservedly popular science', especially as practised by Lyell. Although they agreed on the marks of a mature science, Herschel and Whewell followed different emphases in classifying the disciplines, with Herschel being more concerned to reveal their 'mutual relation and dependency' (1830, 144–7, 94).

The most striking feature of Whewell's review was its silence on the details of Herschel's explanation of '*inductive logic*'. He regarded the *Discourse* as the first attempt since Bacon to deliver a connected body of methodological rules, but said little about the way in which Herschel modified or reinterpreted Bacon. As Laudan and others have noticed, Herschel gave far more support to the role of hypotheses than most recent or contemporary writers, even suggesting that the source of conjectures was irrelevant if they were subsequently shown to explain phenomena separate from those which originally suggested them (Herschel 1830, 164–70, 197, 203). This was a significant move from the previous orthodoxy, maintained by authors such as Reid, which rejected any hypothesis not grounded in cautious induction. There was also tension between this latitude with respect to hypotheses and Herschel's apparent support for the Newtonian principle of *verae causae*, since theories such as the undulatory theory of light awarded causal status to unobservable entities (1830, 144, 203, 152–3; Laudan 1981, 129–31).

Whewell subsequently developed this position in his major works, where he also made strong criticisms of the anti-hypothetical attitudes of most Baconians. In the review, however, he did not offer any support to Herschel's liberalism; indeed, he questioned the lack of attention to Bacon's warning about the 'method of *anticipation*'. In discussing John Dalton's atomic theory, Herschel said that it had been announced by its discoverer 'on the contemplation of a few instances, without passing through the subordinate stages of painful inductive ascent'. Whewell feared that this passage might be 'liable to misinterpretation': it might promote the notion of scientific discovery as the result of unprepared genius. To counter this idea he cited Sedgwick – a living member of the 'school of Bacon' – who had just told the Geological Society that 'the records of mankind offer no single instances of any great physical truth anticipated by mere guesses and conjectures'. Similarly, Whewell appealed to the lessons of the past in order to emphasize the danger of 'anticipation', claiming that 'the usual history of physical theory has been the reign of one anticipation after an other in unbroken succession' (Whewell 1831a, 399–401).

It is thus possible to see that reviewing the *Discourse* was not a simple task for Whewell, even though he was well placed to explain and advertise its meaning. He saw that the pages of the *Quarterly Review* provided the chance to 'get *the people* into a right way of thinking about induction' (Whewell to Jones, February 1831, in Todhunter 1876, II, 115). But the task took him two weeks to complete, and he told Jones that he did not 'much like the review of Herschel now that I look at it in cold blood, and I have a strong persuasion that all the philosophical part will repel most and puzzle the rest' (Whewell to Jones, 15 July 1831 in Todhunter 1876, II, 117, 123). The problem here was one of trying to put forward a clear view of the process of scientific investigation when the crucial terms involved were being used in a variety of ways. Whewell and Jones had come to this diagnosis of the situation and welcomed the *Discourse* as a major step towards clarifying the nature of science for a wider audience. But it was difficult for Whewell to engage with the details of Herschel's position in a public forum without undermining this, even though Jones advised that his review would be more 'popular' if he could 'cast about and *abuse* somebody' (Jones to Whewell, 4 March 1831, WP, Add. MS. c. 52[25]).

Consequently, some of the novel features of the *Discourse* – as seen by later writers – were not fully noticed in the review. While it is important not to read Whewell's published views of 1837 or 1840 into his position in 1831, it is likely that he agreed with Herschel's departure from the reigning Baconian orthodoxy on hypotheses. Indeed, two years later, in his address to the British Association, he went further than Herschel in criticizing extreme 'Baconian' empiricism by asserting the controlling role of theory in observation (Whewell 1833a, xx; Yeo 1985, 267, 274). However, any consideration of this would have clashed with his attempt to stabilize the meaning of Baconian terms and the general validity of this account of induction. Whewell also chose not to comment on Herschel's 'admirable precepts and maxims' – his twelve rules – but told Jones that there was still a need for more, hopefully from Herschel, on 'the practical rules and cautions for making experiments and collecting laws from these' (Whewell to Jones, 15 July 1831, in Todhunter 1876, II, 124). It is possible that Whewell was anxious about reducing the method of discovery to easily grasped examples, such as the case of Wells's investigation of dew – used by Herschel – and he hinted at this general problem in the review. When he read Mill's *System of Logic*, which borrowed this illustration from Herschel, Whewell suggested that Jones should 'tell Herschel he has something to answer for, in persuading people that they could so completely understand the process of discovery from a single example' (Whewell to Jones, 7 April 1843, in Todhunter 1876, II, 314).

Perhaps more indicative of Whewell's trouble in the review of 1831 is his later remark to Herschel, also by way of comment on Mill's work:

There is in new books of this kind a satisfaction in which both you and I may share. I mean that notions and expressions, which were new and strange when we began to write, are now familiarly referred to as part of the uncontested truth of the matter. (Whewell to Herschel, 8 April 1843 in Todhunter 1876, II, 315)

Thus, in spite of the differences between Mill, Herschel, and himself on epistemology, Whewell felt that some progress had been made since the early 1830s when even the basic terms employed in the discussion of scientific method were insecure. In this sense, Mill's book, despite its critique of Whewell's idealism, was part of a shared discourse – one that did not exist earlier. Whewell's problem in 1831 was that he had to review the *Discourse* for a general audience before the meaning of terms such as induction, deduction, experiment, hypothesis, fact, theory, and verification had been firmly illustrated with reference to practical examples of successful science. This is what the *Discourse* achieved. But, furthermore, Whewell was caught in the situation of disagreeing with Herschel at a number of points while trying to promote a consensus on the meaning of inductive method for the public.

Lyell and geology

The review of Lyell's *Principles* was Whewell's chance to speak publicly on the most popular science of the day. He pointed, however, to an anomaly. Although this was a time when publicity and popular circulation were 'the criterion of all progress', English geologists showed a 'strong suspicion of all generalities', and hence it had been impossible for 'mere popular readers' to acquire any distinct view of recent work in geology (Whewell 1831c, 180-3). While this partly reflected their assiduous collection of data, which had produced splendid results, Whewell feared that this approach did not assist 'common readers'. 'Such persons', he declared, 'require general propositions, because those are the only ones which to them have the air of knowledge.' The public sought larger views, not minute facts. The problem was that works which offered these 'theoretical speculations on the general points' were hopelessly uninformed because they were not written by 'any good working geologist' (1831c, 183-4). His plea therefore was for specialists to generalize for the education of a wider audience. This is what he set out to do in both reviews of Lyell, later telling his sister that he hoped the one in the *Quarterly Review* 'was not very unintelligible, as it was intended for general readers' (Whewell to his sister, 13 March 1832, in Stair-Douglas 1881, 144).

Against this background, Whewell introduced Lyell as a man for the times, a philosopher capable of casting a judicious theoretical eye on the large quantity of facts now assembled by geologists (Whewell 1831c, 184). Lyell was, he said, a new exponent of the Huttonian doctrine about the efficacy of present forces, over huge periods of time, as causes of all past geological phenomena. Indeed he was bolder than Hutton, who occasionally had recourse to something more than 'common volcanic eruptions': 'Mr. Lyell throws away all such crutches; he walks alone in the path of his speculations; he requires no paroxysms, no extraordinary periods; he is content to take burning mountains as he finds them' (1831c, 185).

Whewell was able to use the review of Lyell as a vehicle for some reflections on the history of geology as a relatively new science. He also did this in the same year for the more specialized audience of the *Edinburgh New Philosophical Journal* in an article on 'The progress of geology'. In both essays Whewell endeavoured to show that geology, although a recent science, was a subject with a history and heroes, one marked by four stages connected with the analysis of four strata. Whewell associated these stages with the contributions of great individuals: Abraham Gottlob Werner, William Smith, and Georges Cuvier. Cuvier brought his 'magical power' to the study of the tertiary period, creating a new science – comparative anatomy – one that revealed a 'succession of different races of animals' and suggested geological 'revolutions which had taken place in a manner and order hitherto unguessed' (Whewell 1831c, 189 and 1831e, 245). The fourth stratum, which Whewell dubbed '*Penultimate*' formations, was that belonging to 'the state of things which last preceded that under which we live and geologize' (Whewell 1831c, 190). No great man of science was yet associated with these, but in noting that Lyell had provided a mass of new facts about them, Whewell suggested that his aspirations were grand. As the review in the *British Critic* unfolded, it became clear that Lyell's pretensions to heroic stature as a geological theorist rested on a vision alternative to Cuvier's revolutions in the history of the earth.

Whewell suggested that Lyell had moved beyond descriptive, or phenomenal, geology to attempt an explanation of the passage from the geology and organic life of the penultimate era to that of the present, to show that apparent discontinuities could be accounted for by natural causes, such as the changing relation of sea to land. Thus fossils now embedded in mountains were formerly creatures living in the sea. He spelt out the strategy behind Lyell's illustrations: if the passage from the *penultimate* to the present state of the earth could be explained by existing causes acting slowly over immense time, then could not the transition from the 'tertiary to the penultimate period' be part of 'the same chain of events'? (Whewell 1831c, 193). Lyell attempted to demonstrate this by asserting the 'uniformity of nature on a great scale', by reconstructing the effects of present geological agents – mainly aqueous and igneous – acting gradually over an expanded timescale (1831c, 202). Whewell doubted that Lyell could convince 'common readers' that the elevation of the Andes from the bed of the Pacific was a phenomenon 'of the same kind' as those observable in modern times. But he admitted that in addressing this question Lyell had begun to profile a separate science of 'geological dynamics' that sought to classify and analyse the continuing changes in the inorganic portion of nature (for elaboration of this concept, see Whewell 1838b, 13–28 and 1857a, III, 450–95).

When he knew the contents of the review, Lyell wrote to his sister saying that: 'Whewell of Cambridge has done me no small service by giving out at his University that I have discovered a new set of powers in Nature which might be termed "Geological Dynamics". He is head tutor at Trinity, and has more influence than any individual, unless it be Sedgwick' (Lyell to his sister, 14 November 1830, in K. M. Lyell 1881, I, 312). However, Whewell argued that Lyell's important ideas on geological dynamics could not support his speculative vision of a 'perpetual recurrence of cycles of change of the same kind' (Whewell 1831c, 202). Lyell charged that former theorists had erred in assuming that the past was governed by rules 'quite different from those now established'; but Whewell replied that Lyell's own supposition of complete uniformity was also questionable. Apart from the fact that the often discontinuous relationships between the various strata suggested catastrophe rather than gradual change, Lyell was wrong to claim logical superiority for his view. What, asked Whewell, do we know of igneous powers which entitles us to assert that they may not have varied in power over time (1831c, 201–4)? On the other hand, if uniformity of causes were granted, theoretical consistency would demand that Lyell provide 'some mode by which we pass from a world filled with one kind of animal forms, to another, ... without perhaps one species in common' (1831c, 194).

In his review of the second volume, in the *Quarterly Review* of March 1832, Whewell made one of his most significant interventions in the geological debate. He coined the terms 'Uniformitarian' and 'Catastrophist' to denote, respectively, Lyell's position and that of his opponents. This allowed him to depict a conflict between two theoretical perspectives and, indeed, to contrast Lyell with Cuvier. Whewell claimed that the Catastrophist position, which embraced Cuvier's notion of revolutionary discontinuities between different geological eras, was the reigning doctrine and the one most consistent with observations such as the 'successive creation of vast numbers of genera and species' (Whewell 1832a, 126). If phenomena such as these – of which there was no observable modern equivalent – had occurred in the organic sphere in the past, it was likely that 'mechanical operations' quite different from those of the present have also occurred, as the Catastrophists believed. In Whewell's view, the only way Lyell could escape this dilemma was to give a naturalistic account of the emergence of new species. Since Lyell had attacked Lamarck's theory of transmutation in his second volume, Whewell knew that this was a major problem; but by exposing this dilemma he was able to show readers that Catastrophism offered an equally grand picture of the creation, one compatible with prevailing religious convictions:

> We conceive it undeniable, ... that we see, in the transition from an earth peopled by one set of animals, to the same earth swarming with entirely new forms of organic life, a distinct manifestation of creative power, transcending the operation of known laws of nature. (Whewell 1831c, 194)

Whewell said that in this way geology had 'lighted a new lamp along the path of natural theology', but in both reviews he sought to affirm the liberty of geologists to investigate the history of the earth, particularly if they did so like Lyell, with 'the dignity of a philosopher' (Whewell 1831c, 205–6). He doubted that 'those who endeavour to fasten their physical theories on the words of scripture are likely to serve the cause either of religion or science' (Whewell 1832a, 117). At this stage he had not explicitly excluded the question of species' origin from the domain of inductive science, as he did in the *History*, but he had underlined the importance of geology as a subject that straddled the boundaries of the physical and moral spheres. Moreover, he asserted the right of *geologists* to advance synthetic views in public (Whewell 1857a, III, 625; Corsi 1988, 147. For an analysis of Whewell's geological vision, see Hodge 1991.)

Political economy

The review of Jones's work, *An essay on the distribution of wealth and on the sources of taxation: Part 1 – rent* of 1831, gave Whewell a chance to write about another new science – political economy. His personal investment in this case, however, was much higher because Jones was his close friend. The appearance of this first part of the book was largely due to Whewell's support – and at times, despair – as Jones painfully laboured his way towards completion. The part which did appear in 1831 was regarded as a joint production. In April 1831 Jones wrote to Whewell in terms that indicate his debt: 'And now my dearly beloved do remember that you have got one foster child in the world and that although it grow but slowly still it promises eventually to do us both some credit.' A month later he sent 'good news of the baby' (Jones to Whewell, 22 April 1831 and 19 May 1831, WP, Add. MS 52[34 and 36]). The second part, on wages, was never completed, although parts of it were published in a posthumous collection of Jones's papers edited by Whewell (Jones 1859).

Why was Whewell, the author of texts on mechanics and the Professor of Mineralogy, concerned with the study of rent, wealth, and taxes? One answer has drawn attention to the manner in which Whewell's own financial interests were bound up with the fortunes of Trinity College as a powerful landowner benefiting from the high domestic prices guaranteed by the Corn Laws. His criticism of David Ricardo's claim that these laws produced high rents might therefore be seen as a defence of personal interests (Checkland 1951, 43–70; Ruse 1991, 104). Whatever the evidence for this – and nothing substantial has been supplied – it is hardly sufficient to explain the degree of his intellectual involvement in the progress of Jones's work.

A close study of their letters from 1821 shows that they both saw the outcome of contemporary debates on political economy as crucial for a proper view of inductive philosophy and science. Their assessment of what was at stake is reflected in Jones's battle plans. His aim was the establishment of 'sound political and moral views on a subject which has hitherto only called itself a science to enforce a dogmatic philosophy of the most pernicious kind – we must not be too bold in talking of the execution' (Jones to Whewell, 27 September 1827, WP, Add. MS. c. 52[15]). Jones was concerned that the reigning doctrines of political economy advanced by Ricardo and his followers had used the term 'science' to legitimate conclusions arrived at by faulty methods. In short, Whewell and Jones argued that the doctrines of the Ricardians could not claim the status of 'science' because they rested on inadequate empirical foundations. In two papers presented to the Cambridge Philosophical Society, Whewell pursued the strategy of converting the principles of Ricardo into mathematical equations and deducing their consequences, showing, in the second paper, that these were at odds with Ricardo's conclusions (Whewell 1830b and 1831f). Whewell thought that this left the way open to challenge the reasoning and the premises of the school. Jones, however, was worried that this tactic might have the effect of endorsing the so-called axioms of Ricardo, since Whewell had, for the purposes of argument, left these unchallenged. He told Whewell to:

add a passage stating that you take the axioms about rent merely hypothetically – in order to shew the conclusions which right reasoning ought to have led those who assume them – you cannot do this too decidedly because in truth the axioms as applied *generally* to rents, are stark naught and only apply as far as they apply at all to a very limited class of rents. (Jones to Whewell, 18 April 1829, WP, Add. MS. c. 52[16])

Jones demonstrated this last point in the volume on *Rent*, and Whewell's review in the *British Critic* of July 1831 made precisely the clarification Jones requested. In summarizing the material of the book Whewell stressed that one of its lessons was that the distribution of wealth takes place 'in different parts of the world according to very different principles and rules' (Whewell 1831d, 42). This applied to the case of rents, which Jones classified into labour, serf, métayer, ryot, and cottier rents, citing examples of these from various countries. With the Ricardians in his sights, Whewell was astonished that 'our modern political economists' should have attempted to found universal principles on the basis of farmers' rents alone. This, he said, indicated the irresponsible manner in which they constructed their system, one which provided the 'most glaring example of the false method of erecting a science which has occurred since the world has had any examples of the true method' (Whewell 1831d, 51–2).

Whewell set out in some detail the kind of errors involved, but the salient point was that Ricardo and his followers had prematurely sought to give political economy a deductive form. Since it had to be founded on the realities of human affairs, it could not hope to deduce its conclusions, like geometry, from a 'few definitions and conventions'. 'It must', Whewell stressed, '*obtain* its principles by reasoning upwards from facts, before it can *apply* them by reasoning downwards from axioms' (Whewell 1831d, 52). It is important to recognize that Whewell was not excluding the possibility of deduction from general principles in political economy; but he noted that only the most mature of the physical sciences had reached this form: most still awaited their Newton. It was therefore unlikely that political economy, a new discipline dealing with the most complex of subjects, could spring at once to the level of mechanics or astronomy. What Whewell called the 'vices of their method' was the Ricardians' unwillingness to give the science an adequate empirical base. In contrast, Jones had begun to do this by collecting and examining evidence 'in different parts of the earth, and at different periods' (1831d, 53, 55–7).

The final point in Whewell's recommendation of Jones's work also depended on a contrast with his opponents. In this case it concerned 'moral tone'. Whewell claimed that the prevailing school of political economy represented man as a mere instrument for the production of wealth, whereas Jones considered economic activity in conjunction with social, moral, and religious factors. Rather than seeing humans as inevitably engaged in conflict with each other and with impersonal forces, Jones argued that the 'common welfare of each class resides in the common advance of all' (Whewell 1831d, 60), and that there was a moral economy designed by God for human progress. Whewell would no doubt have liked to link the methodological vices of the Ricardians with the low moral view of man taken by the utilitarians who embraced their political economy. As we shall see in later chapters, he believed that there was such an association between correct scientific method and sound moral views. However, in this case the situation was complicated by the realization that Oxford academics, such as Richard Whately and Nassau Senior, who could not easily be charged with irreligious motivations, also committed these serious errors of method.

Corsi has shown that Jones and Whewell became aware of this in April 1831 and that Whewell, in particular, made several indirect assaults on their position. He tried one of these in the review of Jones, quoting without reference a passage by Senior which had been published in Whately's treatise on logic of 1829 (Corsi 1987, 125). Senior had expressed surprise that there was any more difference of opinion among political economists than among mathematicians, since the subject was founded on 'a few general propositions deduced from observation or from consciousness, and generally admitted as soon as stated' (Whewell 1831d, 56). Whewell was appalled at the Oxford academics' endorsement of Ricardo's error of method: namely, the delusion that general principles could be obtained by 'some transient and cursory reference to a few facts of observation or of consciousness' (1831d, 53).

Both Whewell and Jones regarded the pronouncements of the Oxford 'logicians' as a severe threat to their own attempt to explain the nature of inductive method to the public. But it was more difficult to name them explicitly, especially in the pages of the *British Critic*, for although Whately was by no means a conformist member of the Anglican establishment, he did become Archbishop of Dublin in 1831. In that year Jones counselled Whewell: 'Do not unnecessarily

provoke the Oxford men – they may be persuaded to work and help –
the Cockneys are past redemption' (Jones to Whewell, 7 March 1831,
WP, c. 52[26]). This underlines the point that the London-based
utilitarians were far removed from the common ground shared by
Whewell, Jones, Whately, and Senior: namely, the affirmation of the
existing nexus between Church and university and the importance of
a Christian framework for knowledge against the growing forces of
secularization. What Whewell arguably wanted to say, but couldn't,
was that the Oxford academics were allowing the Ricardians and the
utilitarians to appropriate the name of 'science' for their false and
morally dangerous doctrines. This kind of difficulty may well explain
why Whewell, who had dampened Rose's long-held enthusiasm for a
new journal as recently as 1830, should revive the idea a year later
with Jones (Whewell to Rose, 10 January 1830 and 24 April 1831, in
Todhunter 1876, II, 106, 118).

Idea of science in the reviews

Some commentary on the image of science conveyed by Whewell in
these reviews is now possible. Perhaps the most striking feature is the
extent to which he stressed the differences within science, showing
that different disciplines were at various stages of development. This
was most explicit in the review of Herschel's *Discourse* where he spoke
of a hierarchy of sciences arranged according to maturity – that is,
judged in terms of mathematical expression of general laws. This
produced an ambiguity in the meaning of the term 'scientific' in
Whewell's essays. For example, the review of Herschel did not present
a positive picture of the state of chemistry or geology, or indeed of any
discipline apart from the physical sciences of astronomy, mechanics,
and optics. Here the message was that these were the only successful
and certain sciences, whereas other subjects were either severely
limited in scope or torn by controversy. On the other hand, it is clear
from the review of Lyell that Whewell did believe that progress was
being made in geology, and that a geological dynamics, following the
model of astronomy, might be possible. Here he used the term
'scientific' in a second sense, referring not to achievements but to
processes, not to demonstrated laws, but to the use of proper method.
   In all these articles he stressed that progress in inductive sciences
was only possible if they followed the method vindicated by the
history of science. In broad terms, this was the Baconian prescription

to observe, collect, and classify data, and only then to attempt generalizations. This also demanded adequate comparative data. Thus in discussing the history of geology, Whewell's praise of Werner was qualified by the limited evidence on which his system had been built:

He had merely explored a small province of Germany, and conceived ... that the whole surface of our planet, and all the mountain chains in the world, ... were made after the model of his own province. (Whewell 1831e, 254)

This was certainly not a criterion invented simply to attack the deductive model of Ricardo's political economy, but it was one that Whewell promoted against other ways of denoting the attributes of a science.

One of these was the notion that the possession of exact definitions was the mark of a science. This had acquired polemical currency in the debates over political economy because the Ricardians made much of their strict definitions of rents, wages, and capital, invoking the comparison with geometry. Whewell wrote an article in the *Philological Museum* in 1833 to refute what he acknowledged as the widespread acceptance of this notion, but he also sought to clarify it for the less specialist readers of the *British Critic*. His first point was that the analogy with mathematics and geometry was inappropriate because in these subjects the definitions were themselves the 'first principles of our reasonings' (Whewell 1833b, 264). However, in the physical sciences dealing with the external world definitions were helpful only if they captured relationships already discovered in nature. Taking examples from mechanics and optics – two successful physical sciences – Whewell argued that the clear definitions of terms in these disciplines *followed* the settlement of a dispute or the establishment of a theory. His conclusion was that 'exact definitions have been, not the causes, but the consequences of an advance in our knowledge'. Thus the improvement in nomenclature in chemistry by Lavoisier was intimately linked with his contributions to theoretical understanding. Whewell indicated that it was absurd to believe that definitions alone would assist further progress in chemistry: a fixed meaning for, say, 'acid' and 'alkali', could not be imposed while there were continuing debates 'among the highest authorities concerning their boundary lines' (Whewell 1833b, 264, 267).

This situation made it all the more ridiculous for the exponents of political economy to parade exact definitions as the proof of its scientific status. But in order to enforce this point Whewell had to undermine the analogies drawn by Ricardo – and the Oxford writers – between mature deductive sciences and political economy. In addition, he had to emphasize the distance between this subject and those dealing with the natural world. It was the complexity of the social realm that necessitated thorough empirical research; until this was done – and in his view it had only just begun – the imposition of fixed definitions was radically premature. Furthermore, there were moral and political values in the issues of political economy. Whewell stressed that, unlike definitions of lines and circles, propositions about rent and taxes, wages and capital, were things which involved passions and interests (Whewell 1831d, 56). All of this had the effect of underlining the distinctions between sciences, thus making it more difficult to present a unified concept of science at a public level.

This tension in Whewell's message was compounded by the fact that his reviews had two audiences – the educated public and the experts whose works were assessed. Whewell often referred to the 'common reader' or the 'daily and hourly widening circle of intelligent readers' and, as I have noted, believed that specialists had a duty to offer accessible accounts of science. If they did not, it was likely that uninformed popularisers would produce misleading appraisals, thus damaging the public understanding of science. But these reviews, and those by other major commentators such as Brewster, were also aimed at the expert community. Thus Herschel thanked Whewell 'for the high treat of your two reviews of the science of the English Universities and Lyell's Geology – the latter especially I have read over till I have almost got it by heart, and think more of it than I will trust myself to say' (Herschel to Whewell, 29 September 1831, in Todhunter 1876, 1, 57 and Jones to Whewell, 20 March 1831, in 1, 61). The second review of volume II of the *Principles* was written at Lyell's request, and he acknowledged that a note from Whewell, saying that he had begun the review, 'excited me to work with spirit' (journal entries for 12 November and 2 December 1831, in Lyell 1881, 1, 351, 355). It is also evident that experts desired a review that was lively and readable and therefore attractive to the journals' readers. On this score, for example, Lyell thought Whewell's second review was good, but 'not so spirited in style' as the first one (Lyell 1881, 1, 359).

On the one hand, this indicates the location of science in a public forum together with subjects such as theology and politics. But at the same time it meant that certain issues, such as the logic of induction, the appropriate analogies between the sciences, and the role of terminology, were discussed in public while they were still matters of debate amongst experts. The original works of Herschel, Lyell, and Jones – pioneering contributions in their respective fields – raised issues such as these and were thus difficult to review for two audiences. In contrast, Mary Somerville's *On the connexion of the physical sciences* of 1834 was a work of deliberate popularization, albeit a demanding one, and Whewell's account of it in the *Quarterly Review* gave him the opportunity to consider the manner in which science could be presented to a wider audience.

Somerville and the popularization of science

This was not one of his best essays, probably owing to the short notice given by Lockhart who, as noted earlier, was desperate for copy. Consequently, much of the review consisted of extracts from Mary Somerville's book, but it nevertheless contained some revealing comments. Whewell opened by suggesting that there were 'two different ways in which *Physical Science* may be made popularly intelligible and interesting'. It could either be presented as substantive content or discussed in terms of its broader relations and connections. The latter option, he admitted, might be somewhat vague, but it was potentially useful, for 'strange as it may seem, it is undoubtedly true, that such general aspects of the processes with which science is concerned may be apprehended by those who comprehend very dimly and obscurely the nature of the processes themselves' (Whewell 1834b, 54–5).

In saying this Whewell defended the aim of Somerville's book but, more generally, explained the task of attempting a 'popular view of the present state of science'. Drawing an analogy with paintings in which the theme and setting are evoked without reliance on detailed representations, Whewell affirmed that 'language may be so employed that it shall present to us science as an extensive and splendid prospect, in which we see the relative positions and bearings of many parts, though we do not trace any portion into exact detail' (1834b, 55). Somerville had accomplished this in the complex realm of the mathematical and physical sciences. He observed that this was an area in which popular ignorance was high, citing the case, mentioned by Somerville, of the alarm in Paris when it was predicted that a comet would pass close to the earth's orbit. François Arago wrote an article to attempt to dispel these fears, explaining that the orbit of the earth was not a material thing that could be damaged by this event. Whewell cited this as an illustration of the 'confusion of ideas' which might be clarified by books such as Somerville's.

Whewell then moved from the question of popularizing scientific laws to that of defining the image of science. Somerville's work provided a bridge to other themes: the unity of science and the place of women in it. The *Connexion of the sciences* allowed Whewell to raise the spectre of the 'separation and dismemberment' of science as the universalists of the past such as Hobbes and Goethe were replaced by specialists indifferent to each other's research: 'The mathematician turns away from the chemist; the chemist from the naturalist; ... the chemist is perhaps a chemist of electro-chemistry; if so, he leaves common chemical analysis to others.' In this way, he concluded, even the physical sciences lose 'all trace of unity'. Whewell interpreted Somerville's work as one possible response to this situation, in that she showed how detached branches have been 'united by the discovery of general principles'; relationships or analogies had been found between different phenomena, such as 'the electric and magnetic influences'. Implicit in her book, he added, was another connection between the various physical sciences: namely, 'the community of that *mathematical* language which they all employ' (Whewell 1834b, 59–60). Clerk Maxwell, remarking later on the large sales of this book, said this showed that 'there already exists a widespread desire to be able to form some notion of physical science as a whole' (Maxwell 1890, II, 401; also Yeo 1986a).

But while this issue of the unity of science was clearly raised by Somerville, Whewell used the review to place it within a wider agenda. Rather than confining himself to the question of specialization in the physical sciences, he contended that the entire scientific enterprise was in danger of losing its integrity. 'A curious illustration of this', he said, was 'the want of any name by which we can designate the students of the knowledge of the material world collectively' (Whewell 1834b, 59). Thus, at the 1833 meeting of the British Association there was 'no general term' by which the students of natural knowledge could describe themselves. Whewell then proceeded to relate the way in which the term 'philosopher' was disallowed by Coleridge as 'too wide and too lofty' and how 'some ingenious gentleman' – that is, himself – proposed the word 'scientist'. In suggesting this term Whewell not only effectively distinguished 'scientists' from 'artists' but sought to highlight the common enterprise in which astronomers, chemists, geologists, and botanists were engaged. This was part of a strategy to prevent the disintegration of what both he and Harcourt called 'the empire or commonwealth of science'. It is significant that Whewell chose to raise this issue, and publicize his neologism, in the periodical press; for in this way he charted out a space for metascientific comment far more ambitious than Somerville claimed.[5]

Whewell devoted a considerable section of the review to a reflection on the fact that the *Connexion* was the work of a woman. In fact, he told Lockhart that he was aiming to give an opinion of 'Mrs. Somerville and her book' (Whewell to Lockhart, 29 January 1834, NLS 924[41]). It is not too speculative to regard this as relevant to the major theme about the unity of science, for Whewell implied that, against common prejudices, science need not be divided by gender. This was not so much because he believed there were no characteristic male or female mental traits. There was, he said, 'a sex in minds': in women, action was the result of feeling, thought of seeing; when females theorized, they did so without being, like men, perturbed by practical applications. But the consequence of this was that 'when women are philosophers, they are likely to be lucid ones' (Whewell 1834b, 65).

The heavily ambiguous message here seemed to be that Somerville showed that a woman can comprehend physical science and so its potential audience was not solely male. In fact Whewell had said this more explicitly about her previous book on *The mechanism of the heavens*. In a new edition of his *Treatise on Dynamics* of 1832 he told a university readership that her work showed that these severer studies were 'reconcilable with all the gentler train of feminine graces and accomplishments' and could no longer be 'represented as inconsistent with a polished taste and a familiar acquaintance with ancient and modern literature' (Whewell 1832e, v). Somerville had addressed the *Connexion* to her fellow 'countrywomen', but Whewell remarked that it contained much that would surprise individuals of 'that gender which plumes itself upon the exclusive possession of exact science'

---

[5] The coining of 'scientist' in this context is missed by authors who follow the *OED* reference to Whewell's later use of it in the *Philosophy*. In France, there was not even the equivalent of 'natural philosopher', since *savant* referred to scholars in general; hence the French had already begun to speak of *mathématicien* and *chemiste*. See Paul 1980, 105.

(Whewell 1834b, 56). The prospect of a large audience, both male and female, for science was attractive to Whewell and other promoters of the scientific enterprise. However, there was nothing in Whewell's praise of Somerville to suggest that women could be creative discoverers as well as discerning expositors of science. Somerville herself certainly felt that the former role was beyond her, and later recalled that: 'I was conscious that I had never made a discovery myself, that I had no originality. I have perseverance and intelligence but no genius, that spark from heaven is not granted to the sex' (Patterson 1969, 318). The great irony here is that Whewell himself also believed that he lacked this capacity.

SCIENCE AND THE CRISIS OF THE PUBLIC SPHERE

After this period of intensive reviewing from 1831 to 1834 Whewell did not write many other articles on science in periodicals, although a review of Comte's *Positive Philosophy* in *Macmillan's Magazine* was one of the last things he published. As mentioned earlier, Whewell had long held doubts about the value of periodicals; and by the time he wrote the *History* he was not satisfied with the role of reviews as guiders of public attitudes on science. He suggested that the bad reception of Thomas Young's theory of light showed that there was 'in England no visible body of men, fitted by their knowledge and character to pronounce judgement on such a question, or to give the proper impulse and bias to public opinion'. The 'writers of "Reviews" alone, self-constituted and secret tribunals, claimed this kind of authority' (Whewell 1857a, II, 346). These comments are in keeping with his conviction that the role of scientific commentator was a serious calling and one that, if abused, could produce damaging results. As late as 1857 he explained to Forbes the danger of British Association presidents and Edinburgh reviewers promoting theories that were still speculative, saying that 'theirs is a judicial position, and they give ordinary readers a wrong notion of the progress of science' (Whewell to Forbes, 19 September 1857, in Todhunter, II, 411). However, his own reviews attempted not only to adjudicate new scientific theories, but to promote the scientific enterprise itself. He remained unsure of their success, recalling in 1856 that 'I have written very few review articles in the course of my life; and those have not been popular, except one or two on scientific subjects' (Whewell to Kate Malcolm, 18 April 1856, in Stair-Douglas 1881, 470).

Against this background it is significant that Whewell did not engage in the flurry of reviewing that followed one of the most public threats to the authority of the scientific community: the appearance of Robert Chambers's anonymous *Vestiges* in 1844. Indeed, Whewell saw the success of this book as an indication of the failure of scientific commentators, and perhaps himself, in explaining good science to wider audiences. Sedgwick did what he saw as his duty in the form of an eighty-five-page onslaught in the *Edinburgh Review*, but Whewell, together with Forbes and Owen, thought a direct reply would only legitimate its claim to a scientific hearing. Whewell later counselled Sedgwick that 'the great difficulty in satisfying the mind of general readers on such subjects is that you have to oppose to attractive positive generalization nothing but negations and doubts' (Whewell to Sedgwick, September 1849, WP, O 15. 48[69]). Napier seemed to appreciate the problem posed by *Vestiges*. He told Whewell that he had asked Sedgwick to review this book because its 'Geological Speculations brought the book more within his chosen range than yours', but then realized that a good article would 'require attainments in Science and Philosophy – though Geology and its connected walks are, perhaps, more particularly concerned – almost as extensive and various as your own' (Napier to Whewell, 8 February 1845, WP, Add. MS. a. 210[10]). Thus Napier, positioned at the intersection of experts and educated readers, could see Whewell's special role. Intriguingly, at least for those outside the elite scientific community, it may have taken a popular, synthetic work like Chambers's to highlight the importance of Whewell's metascientific criticism. However, he declined the offer to review.

Instead, Whewell republished sections of his major works in 1845 as *Indications of the creator* with a preface charging that the author of *Vestiges* had committed a serious intellectual and moral error. In the case of the 'nebular hypothesis' – a term coined by Whewell in his Bridgewater treatise – Chambers had removed a tentative scientific hypothesis from the safe circle of scientific experts and proclaimed it 'to the world' as a dogmatic doctrine (Whewell 1846a, 25). Ironically, there is a sense in which Chambers was performing what Whewell had earlier recommended as an important function: namely, the production of interesting generalizations for the public. The failure of specialist scientists to provide this was precisely the point made by Chambers in justifying his work (Chambers 1846, 179; Yeo 1984). In discussing the matter with Frederick Myers, who was impressed by the synthetic sweep of *Vestiges*, Whewell seemed to admit defeat:

If the mere combining chemistry, geology, physiology, and the like, into a nominal system, while you violate the principles of each at every step of your hypothesis, be held to be a philosophical merit, because the speculator is seeking a wider law than gravitation, I do not see what we, whose admiration of the discovery of gravitation arises from its truth, and the soundness of every step to the truth, have to do, except seek another audience. (Whewell to Myers, 16 March 1845, in Stair-Douglas 1881, 318)

Brewster interpreted the *Vestiges* controversy as a threat to the 'holy alliance' of science and religion guaranteed by natural theology (Brewster 1845, 471; Yeo 1984, 11). But from the perspective taken in chapter 2 it was also the culmination of the tensions associated with the discussion of science in the public sphere. This chapter has shown how the major Victorian periodicals provided a delicate cultural space in which men of science reviewed each other's work, but did so in a *genre* – the essay review – that entailed both opportunities and dangers. Whewell's early reviews are particularly interesting because they were not confined to a treatment of substantive scientific theories; rather, they aimed at metascientific reflection, at general commentaries on the nature of science, its method, its terminology, the relationship between its various disciplines, the possibility of a social science, and the competing perceptions of unity and specializ-ation. That Whewell was prepared to raise such issues in this public forum indicates the resonance they had beyond expert circles; but the dispute created by Chambers underlines the difficulties inherent in this activity.

Scientific reviews in periodicals always had a dual audience, since books were being assessed by peers and also explained to a wider lay readership. Chambers contended that he did not discount any of the 'truths already established' by experts, but wove these into the 'generalization' that this audience was not receiving (Chambers 1846, 404; Yeo 1984). In one sense, then, his argument was not with substantive science at the level of papers and specialist books, but with the broader commentary in the periodical literature on the aims and methods of science and the wider meaning of its findings – in short, with the metascience of men like Whewell.

Nevertheless, by the mid 1830s Whewell's reputation as an adjudicator of scientific developments, at least in the eyes of his peers, had been consolidated by his reviews, by the reports for the British

Association, by his contributions to scientific terminology and, in another way, by his treatment of the religious and moral aspects of science in his Bridgewater treatise on *Astronomy and general physics*. In the next chapter I examine this work as one instance of the way in which questions about the moral character of scientists figured in early-nineteenth-century debates about the nature of science.

CHAPTER 5

Moral scientists

I never heard of any mathematician who was a sceptic in astronomy or physics; and yet, there are few branches of knowledge which lie more open to metaphysical quibbles.

Dugald Stewart (1854–60) [1827], III, 207

Dined with Whewell to meet Arago ... [who] told us anecdotes of Laplace ... [including one about his dismissal of any hypothesis of a Deity]. LaPlace called on Arago begging him to omit this anecdote as the *phrase* might do him hurt, he wd not however deny it.

Romilly's Cambridge Diary, Friday 26 September 1834,
Bury 1967, 61

One mark of the consolidation which had taken place in the status of science during the nineteenth century was the ability of Francis Galton to single out English men of science as a distinctive group for psychological study. In 1874 Galton inquired into the nature and nurture of this group and felt reasonably confident that he had isolated a significant cluster of attitudes and aptitudes common to such individuals. He did not, however, 'attempt to define a "scientific man"' since, as he said, such groups always blend into others. Thus the traces of the earlier problematic profile of men of science remained: Galton remarked, for example, that: 'Some of my readers may feel surprise that so many as 300 persons are to be found in the United Kingdom who deserve the title of scientific men' (Galton 1874, 6–7).

As suggested in chapter 2, it was in part this relatively indefinite social profile of the man of science that made a promotion of the values of science so essential. But it also meant that there was a special focus on the type of person who chose such an uncertain career. Well before Galton's research, biographies of men of science were a crucial medium for the presentation of this scientific character, as a means both of showing its conformity with existing models of virtuous behaviour and for explicating its distinctive features (Theerman 1985 and Yeo 1988). In 1896 Merz remarked that 'the history of English science during the first half of the century consists of a series of biographies, or of monographs on single ideas and points of view' (Merz 1896–1904, I, 278). He related this to the weaker institutional base for science in England compared with that in France or Germany. Whereas in these countries debate was conducted by schools, 'in England alone the person of the thinker has nearly always claimed the greater share of popular attention' (1896–1904, 278–9).

It would be wrong, however, to see this emphasis on individual scientists as merely a reflection of the absence of strong institutions. Morrell, for example, has argued that this 'individualism' of British science was also related to other cultural assumptions about self-help (Morrell 1971b). Furthermore, concern was not only with individual freedom but with the moral character of students of nature, and this in turn related to the fact that moral qualities and responsibilities were seen to inhere in individuals, not institutions. There was of course a counter theme to this: namely, the growing stress on more reified conceptions of the distinctive features of science, such as its method (Schuster and Yeo 1986). The culmination of this trend was the notion that truth in science was achieved by the application of codifiable procedures that circumvented the emotions, interests, and biases of the individual scientist. Robert Merton referred to this when he dismissed the view that the infrequency of fraud in science could be attributed to the 'moral integrity' of scientists, suggesting instead that the social norms of the scientific community were influential (Merton 1973, 276). But during the early Victorian period the person, as much as the scientific process, was a source of authority, and the tensions between these competing images of science were still being contested.

Whewell did not write a full-scale biography, but the issues pertinent to this genre were present in some of his major writings. In early inductive notebooks he referred to 'the person associated' with the first step on the path of discovery; and he noted that histories of science written by Cambridge men for the Society for the Diffusion of Useful Knowledge exemplified 'the best scientific history, the intellectual biography of great men' (WP, R 18. 17[5(1)]; Whewell 1831b, 88). This view of science as dependent on the contributions of heroic figures was apparent in the *History*, where men such as Kepler, Newton, and Fresnel were lauded. But although some correspondents thought Whewell was working on a biography of Newton, he did not produce a separate work of scientific biography.[1] One reviewer, commenting on a section of his *On the philosophy of discovery*, said that its account of great scientific and philosophical personalities 'does not make them live' (anon. 1860, 366).

But although in the strict sense scientific biography was one of the few literary forms at which Whewell did not try his hand, he was concerned with one of the key issues informing most contemporary examples of this genre: the relation between intellectual and moral character. This chapter delineates the ways in which a concern with the character of men of science informed discussions about the nature of science, such as the connections with natural theology, the moral responsibility of individuals and institutions, the division of labour between theorists and observers, and the ethos of scientific inquiry. It focuses mainly on the period 1831–7, and discerns a common set of issues in debates concerning the British Association, Whewell's Bridgewater treatise, and the interpretation of the dispute between Newton and Flamsteed.

NATURAL THEOLOGY AND MORAL SCIENTISTS

It is well known that Whewell made significant contributions to the subject of natural theology. *Astronomy and general physics*, a volume in the Bridgewater treatises, was his first book on scientific issues for an audience beyond Cambridge students and teachers. (Another work of general interest, on the architecture of German Churches, was published in 1830.) It was the first of the series to appear, in 1833, and the most popular, reaching a seventh edition in 1864. It is usually discussed in terms of its use of physical science in the design argument and its picture of a world held back from the brink of catastrophe by the power and goodness of a Divine Designer (Glisserman 1975). This aspect was certainly important for Whewell's reputation, but his close friends and critics regarded the section on the different moral consequences of inductive and deductive habits of thought as the

---

[1] On Whewell's knowledge of Newtoniana, see Baily to Peacock, 11 January 1834, WP, Add. MS. a. 200[206]; and Macaulay to Whewell, 2 December 1837, WP, Add. MS. a. 209[152]. For his inquiry about Newton papers in Oxford, see Whewell to Mrs Buckland, 10 October 1831, RS MS. 252[43]. In about 1821 he recorded that the Newton manuscripts in the Portsmouth collection contained many notes on alchemical writings (WP, R 18. 9[9]). Did this dissuade him from a biography, just as it later troubled Brewster? See Yeo 1988, 270–8.

most novel and controversial. Rose declared that 'your distinction between the Discoverers and the mere Mathematicians' was 'as true as truth itself' (Rose to Whewell, 1833, WP, Add. MS. a. 211[143]).[2]

This theme was not entirely foreign to the tradition of natural theology which, in the broadest sense, claimed that natural knowledge had moral benefits for its students. Appreciation of God's greatness as revealed in nature, his 'second book', was the most often cited, but some writers were more specific. For example, both Herschel and Brougham, in two important apologias for science, asserted that scientific inquiry produced a moral effect on the minds of its cultivators. In his *Discourse*, Herschel suggested that the great discoverers, by comprehending the works of the Creator, elevated the dignity of their fellow mortals by allowing them to share in this relationship (Herschel 1830, 16–17). These claims made the status of science closely dependent on the character of the persons who embodied it. In his *A discourse of the objects, advantages, and pleasures of science* of 1827, Brougham had taken this to its extreme:

It is surely no mean reward of our labour to become acquainted with the prodigious genius of those who have most exalted the nature of man above its destined sphere; and, ... to know how it comes to pass that by universal consent they hold a station apart, rising over all the Great Teachers of mankind, and spoken reverently, as if NEWTON and LAPLACE were not the names of mortal men. (Brougham 1827a, 47)

This elicited a hostile response from some High Church theologians at Oxford, the Tractarians, who feared the growth of a secular morality founded on science. In 1834 William Sewell warned that 'a nation of Newtons could no more produce a gentleman than a nation of infidels could create a Christian' (quoted in Corsi 1988, 123). John Henry Newman observed that Peel had also adopted this notion of science as a form of moral elevation, speaking of Newton and Laplace as 'if they are not mortal men' (Newman 1841, 6).

Thus when Whewell came to write a work of natural theology, the moral influence of science was already a contentious issue. In addressing this issue he responded to the charge that some famous men of science were irreligious, and did so in a way that made

---

[2] Whewell said: 'I want most to know what impression is made by the part containing the contrast of induction and deduction' (Whewell to Jones, 24 March 1833, in Todhunter 1876, II, 161). In 1827 Stewart had undertaken a similar analysis of the 'varieties of intellectual character' of metaphysicians, mathematicians, and poets. See Stewart 1854–60, IV, 185–249. On Whewell's design argument, see Brooke 1991a.

questions of character central to debates about the nature of science. In reviewing Herschel's *Discourse*, Whewell described it as one of the best treatments of the 'moral conduct of intellect', citing the passage where Herschel replied to the charge that 'natural philosophy' and indeed, 'all science', fostered self-conceit and religious indifference (Whewell 1831a, 400, 406; Herschel 1830, 7). In opposition to this Herschel proposed that science opened the 'well-constituted mind' to all truth, and taught the philosopher that 'humility of pretension, no less than confidence of hope, is what best becomes his character'. Whewell seemed to reverse this conventional answer, arguing that if men of science were virtuous then so would be their science.

WHEWELL'S BRIDGEWATER TREATISE: MORAL MEN AND MORAL METHODS

In Book 3 of the Bridgewater treatise, headed 'Religious Views', Whewell devoted chapters 5 and 6 to the moral and religious implications of inductive and deductive habits of mind. This discussion followed his claim that the evidence of design, order, and law in the material world produced in man the conviction of an 'intelligent and conscious Deity' as the author of the universe (Whewell 1834a, 301–2). The 'object of physical science' was to discover the laws and properties of this designed world. After urging that scientific understanding of nature enhanced the belief in an intelligent Deity, Whewell sought to show that religious belief was not inconsistent with the character of the greatest men of science.

The problem here was the characterization of these men. Whewell affirmed that they were the 'great discoverers' who had detected the comprehensive laws of nature; but since very few of these laws had been disclosed, the number of such names was limited. In seeking to 'analyse the process of thought' by which these minds attained such knowledge, he admitted that the task was therefore formidable, since there were 'so few instances successfully performed'. Nevertheless, Whewell proposed that the mental habits of such special individuals encouraged the belief in a benevolent Creator. The reason for this, he argued, could be seen if we traced the intellectual path of inductive discoverers. Faced with a mass of facts which seemed incoherent, such inquirers struggled to discern some form of order. Thus Kepler wrestled with the known periodic times and diameters of the planetary orbits until suddenly, in Whewell's words, previously disconnected observations assumed 'the aspect of connexion and intelligible order' (Whewell 1834a, 304–5). In the most celebrated discovery of all science, Newton perceived the relationship between Kepler's law and 'all other known properties of the solar system'. The crucial element here for Whewell was the nature of this transition from the disorder of facts to the perception of a law:

This step so much resembles the mode in which one intelligent being understands and apprehends the conceptions of another, that we cannot be surprised if those persons in whose minds such a process has taken place, have been most ready to acknowledge the existence and operation of a superintending intelligence. (Whewell 1834a, 307)

After giving miniature accounts of the achievements of Galileo, Copernicus, Kepler, and Newton, Whewell stressed their piety and the manner in which it was strengthened by their science (Whewell 1834a, 309–17). Unfortunately, there was a complication here: he admitted that the 'sciences of experiment do not conduct so obviously as sciences of observation to the impression of a Divine Legislator of the material world'. Nevertheless it was possible to mention Boyle, Dalton, and Black as experimenters who recognized design in nature.

From this position Whewell was able to concede what the High Church critics contended: that there were natural philosophers who displayed a lack of religious conviction. But, he contested, none of these was a great discoverer. This point assumed a strong distinction between the original discovery of laws and the explication of their consequences and applications. The former involved '*inductive* ascent' and the latter '*deductive reasoning*', and these operations implied 'different mental habits' (Whewell 1834a, 324, 326). In broad terms, Whewell described deduction as the task of verifying and developing the general laws outlined by the minority of significant discoverers. It was the 'comparatively humble office' occupied by the majority of those engaged in scientific practice (1834a, 304). It must be remarked here that this was a curious way of bifurcating the scientific community, especially since Whewell was currently involved in debates about the division of labour between observers and theorists – a more obvious Baconian dichotomy, and one in current usage in the British Association. He was forced into the contrast between inductive and deductive thinking by the apologetic requirements of his case: namely, the need to repel the suggestion that physical science was inimical to religious belief. By separating two distinct modes of thought, he was able to suggest that they engendered different moral and religious attitudes. This was a departure from his sermons of 1827, where he replied to the alleged dangers of 'the modern physical and moral sciences', but did not contrast induction and deduction (WP, R 6. 17^{14}; also R 6. 17^{13}).

After describing deductive analysis as a relatively humble office, Whewell listed some of the most illustrious names in eighteenth-century natural philosophy as its exponents. Men such as D'Alembert, Euler, Laplace, and Lagrange, he said, had brilliantly unfolded the ramifications of Newton's theory, but their work was the product of 'mathematical talent' and, however rare this was, it was not the 'genius which divines the general laws of nature' (Whewell 1834a, 329). It was precisely because these deductive reasoners did not engage in the painful struggle to delineate general principles that they were sometimes indifferent to religious impressions. Dealing always with generalizations of the highest level, such mathematicians often regarded the laws of motion and gravitation as 'self-evident and certain *a priori*, like the truths of geometry'. Never having to consider the possibility of these laws 'being other than we find them to be', they overlooked any evidence of their being ordained and hence 'substitute[d] for the Deity certain axioms and first principles, as the cause of all' (1834a, 331–2).

Whewell extended this diagnosis in a significant way. Having illustrated the effects of deductive habits in science, he claimed that they could hinder the recognition of other means of reaching knowledge. Interestingly, given his dispute with Oxford academics over scientific method, Whewell still found it possible to use Whately's *Logic* for his own purposes, noting its claim that deduction developed and explicated those truths already contained in the principles which it took as a premise; it was not a method for generating new truth. Induction, on the other hand, was the method for attaining new knowledge in science and in 'the general course of human experience'. The danger was that deductive thinkers would fail to appreciate the value of speculation in morals and politics, where mathematical precision was inappropriate (Whewell 1834a, 334–7).

The implication of Whewell's argument was extraordinary. In a book intended to affirm the religious value of science he had divided the scientific community into two kinds of thinkers, with the deductive type, the majority, possessing mental habits that impoverished their religious feeling and their ability to appreciate moral evidence and poetic beauty. Whewell suggested that uneven cultivation of the faculties in such philosophers, made them more liable than 'common men ... to miss the roads to truths of extreme consequence' (Whewell 1834a, 338). Although he sought to balance this with a paean to the elite group of inductive discoverers who displayed both intellectual and moral virtue, Whewell made it very difficult to present men of science as unambiguous moral exemplars. As a critic of scientific pretension, Rose could hardly conceal his wonder: 'the way in which you speak of the non-importance of mere mathematicians, coming from a person of your station in science is indeed very important' (Rose to Whewell, 1833 WP, Add. MS. 211[143]).

In his unsolicited *Ninth Bridgewater treatise*, Babbage expressed astonishment at Whewell's views, observing that they lent support to those who regard the pursuit of science as unfavourable to religion (Babbage 1838, iv–v, x). It was this part of Whewell's treatise which caused him to respond on behalf of 'other cultivators of the more abstract branches of mathematical science'. Babbage observed that Whewell seemed to bring this charge of unreligious attitude not only against those who pursued branches of abstract science dealing with the properties of pure number, but 'to all who cultivate deductive processes of reasoning' (Babbage 1838, x–xii). Rather than becoming engaged in an historical argument about the moral stance of particular men of science, Babbage tried to disrupt the nexus Whewell was seeking to make between intellectual styles and moral attitudes. Who, he asked, would ever dream of 'inquiring into the moral or intellectual character of Euclid or Archimedes, for the purpose of confirming or invalidating his belief in their conclusions' (1838, xiii)?

Whewell made it clear that he believed the best arguments in support of religion were founded not on mathematical laws of nature, but rather on the connections between different *kinds* of laws.[3] Although he did not repeat the distinction between inductive and deductive thinking, referring more generally to 'mathematical and physical reasoning', Whewell insisted, against Babbage, that an inquiry into the 'intellectual and moral character' of men of science *was* essential. Presumably Whewell had conducted such an analysis

---

[3] Whewell to Babbage, printed letter, 30 May 1837, WP, 266 c. 80. 149[16]. Hare asked for a 'copy of your answer to Babbage's hateful book' (Hare to Whewell, December 1837, WP, a. 206[170]).

before exempting the Irish mathematician, William Hamilton, from the charge against deductive mathematicians. In any case, the fact that Hamilton could write sonnets carried weight, and Whewell asked Jones whether it would be proper to print one of these in the Bridgewater treatise (Whewell to Jones, 2 February 1833, in Todhunter 1876, II, 154; Whewell to Hamilton, 18 March 1833, in Graves 1882–5, I, 162; also 554, 559).

Whewell's Bridgewater treatise has been rightly seen as a contribution to the apologetics of natural theology. However, the sections just discussed – which Whewell and others thought most significant – shifted the emphasis from science and the design argument to the moral character of scientists. But while this might have been a new move within natural theology, it is possible to recognize this concern with the image and character of the man of science in debates about scientific practice, values, and institutions in the 1830s.

INDIVIDUALS VERSUS INSTITUTIONS

Almost within a single year, the British scientific community lost three of its leading members: Humphry Davy, William Wollaston, and Thomas Young died between 1828 and 1829. It is significant that Harcourt, as the first Secretary of the British Association, mentioned these names and underlined the crucial role of individuals in a small community (Harcourt 1831, 22).[4] Yet, at the same time, this new scientific institution was posing a different view of scientific enterprise – one in which, to be sure, individuals would be important, but also one that emphasized shared norms, cooperation, and group consensus.

Critics of the Association detected a plan to advertise a unanimity of purpose. Dean William Cockburn complained in 1845 that 'the members of the British Association have always been accustomed to act in strict unison. They discountenance all difference of opinion' (W. Cockburn 1845, 34). This, he admitted, made the scientific community more difficult to attack. However, as the leaders of the Association knew, this strategy depended on the exclusion of social and moral questions upon which individuals might differ and in which no norms of reasoning existed. Referring in 1833 to the unexpected formation of a statistical section, Whewell spoke of it as an

---

[4] On the possible fragility of research projects associated with key individuals in French science, and the importance of disciples, see Outram 1980, 34–6.

irregularity and reminded the audience that the 'sciences of morals and politics' concerned the 'passions, affections and feelings of our moral nature'. Such inquiries could only be admitted if they were limited to 'matters of fact' (Whewell 1833, xxviii; Morrell and Thackray 1981, 291–2; also Whewell 1841c, xxxiii).

Various commentators debated this agenda. Managers such as Murchison believed the content of meetings had to be carefully monitored and told Harcourt that 'we should give as little rise as possible to subjects which experience has taught us have often been attended with ridicule: I mean general evening lectures and *exhibitions*' (Murchison to Harcourt, 6 April 1938, in Morrell and Thackray 1984, 304–5). De Morgan, a less involved observer, reviewed the first meetings and suggested that not only was there a need for agreement on what should be discussed, but for a consensus on 'principles and modes of reasoning to which all can appeal without fear of lengthened controversy, or of the creation of party virulence'. That is, the need to avoid unmanageable controversy extended into legitimate matters of science. He hoped this was possible because 'most philosophical questions now in dispute have been reduced to circumstances in which violent collision is hardly possible'. If men of science were to pursue 'what we may call a professional object', this depended not on any 'superiority of human nature in our day', but on the strict confinement of debate within those sciences 'which afford no great aliment to angry discussion' (De Morgan 1835, 154–5). Later he reiterated that: 'Nothing is of so much importance to a scientific society as quiet' (De Morgan 1842, 438).

At stake here were competing claims between individual and institutional identity and values. While the British Association was trying to avoid such conflict, the campaign for the reform of the Royal Society produced explicit focus on the role of individuals within scientific societies. In his *Reflections on the Decline of Science*, Babbage invited readers to judge his comments alongside his own character, which he hoped was sufficiently well known in the 'diminutive ... world of science'. One of his contentions was that the adoption of a unitary institutional persona supported the protection of culpable individuals. Referring to the 'cry against personality, which has been lately set up to prevent all inquiry into matters of scientific misgovernment', he asserted that 'the public character of every public servant' is a legitimate subject of discussion (Babbage 1830, x–xi; De Morgan 1842, 436; also Miller 1983). At a time of intense debate over the Reform Bill, similar issues were testing the scientific community. In what sense did scientific institutions represent the views of individual members? Were the British Association and the Royal Society private or public bodies? What were their social and moral responsibilities? (See Brougham 1827b; Macleod 1983.)

TRACTARIAN CRITIQUE OF INSTITUTIONAL SCIENCE

While the leaders of the British Association were struggling with the proper relationship between public and private responsibility, some of its most virulent critics made this a key point of their assault. The Tractarians had several objections to the Association: for example, its threat as a secular Church, preaching science and philosophy rather than religion as a bond of association, and its restriction of science to physical inquiry (Bowden, 1839, 18–19; Morrell and Thackray 1981, 231–3). But another level of their critique, not sufficiently noticed, was an objection to the assimilation of individual identity into that of a scientific institution. In a major article commissioned by Newman for the *British Critic*, William Bowden attacked the institutional mode of producing knowledge and reasserted the value of individually governed inquiry. He did not fear the progress of science, but like all human pursuits, it had 'its own attendant moral dangers', and these, he believed, were exaggerated by the modern organization of science.

In these days of combination, of co-operation, of joint stock company proceedings, the business of natural science, like all other businesses, must, as a matter of course, be carried on by bodies rather than by individuals. (Bowden 1839, 14)

The contrast evoked was that of the 'solitary student', such as Friar Roger Bacon, working alone for years to produce an *opus majorem*. Instead, in the recent style of scientific pursuit, there was a recruitment of many smaller minds into a programme that sought to table results on a daily basis. There seemed to be two concerns here. First there was anxiety about the way in which unconfirmed or controversial findings would be displayed in public forums, rather than confined to the private discourse of scholars. In the previous year Cockburn, the critic of the geology of Buckland and Sedgwick, made this point in *A remonstrance ... upon the dangers of peripatetic philosophy*, which he addressed to the Duke of Northumberland, the President of the British Association when it met in Newcastle. Cockburn alleged that this body had deserted the safe modes of solitary scholarship and the established universities 'where the sons of science are permanently resident', and was instead trying to ascertain truth in 'the confusion of a mixed and multitudinous assembly' (W. Cockburn 1838, 5–6, 24).

Second, Bowden warned that great individuals could be restrained by the 'dwarfish proportions' of the majority. Scientific advance could, he claimed, be hampered by this, since it was less likely that institutional research 'should originate or admit of reasoning opposed to the current ideas and prejudices ... of the day, than one isolated philosopher, acting for himself, should have the courage or penetration to do so' (Bowden 1839, 16–17). In the light of this position, it is interesting that in 1827 Brougham cast doubts on the advisability of British societies following the French tradition of *éloges* on deceased savants, citing their tendency to personalize issues (Brougham 1827b, 352–65). But this is exactly what the Tractarian critics espoused as a necessity, partly because there were moral issues for individuals to confront, and partly because they feared the group persona of the modern scientific institution.

Thus precisely at the time of Whewell's Bridgewater treatise, there were wider ramifications of the relationship between individuals and scientific knowledge. The debates surrounding the establishment of the British Association raised questions about the contrast between individual and institutional scientific practice. The managers of the Association hoped that exclusion of moral and social issues most likely to divide individuals would preserve a unanimous voice for science. De Morgan even thought that a version of this might have to apply to scientific debate itself. On the other hand, the current criticisms of the unreformed Royal Society highlighted the individual identity of its members. The most hostile opponents of scientific institutions, the Tractarians, condemned the manner in which individual achievement and responsibility might be diluted within a public body.

THE MORAL ROLE OF THEORISTS

In fact, the issues of individual character and authority entered the *internal* debates of the British Association. The Baconian legacy, respected by its leaders, advocated a division of scientific labour. This had a double appeal: it encouraged wide participation of amateur cultivators, and it could be used to justify a role for elite theorists, or 'master spirits', as Harcourt called them (Hankins 1980, 139). The division of labour was strongly advocated by Harcourt as the way in which the activities of provincial amateurs could be supported and, just before the foundation of the Association, Herschel had sanctioned this as particularly appropriate in new disciplines such as geology. Whewell had no objection to this policy if there were proper theoretical supervision – his own research on tides depended on such an arrangement. But in his address at the Cambridge meeting of 1833 Whewell interrogated the philosophical assumptions of over-zealous Baconians. He rejected the assertion that '*facts* alone are valuable in science' by claiming that the antithesis between theory and facts was misleading: 'for it is only through some view or other of the *connexion* and *relation* of facts, that we know what circumstances we ought to notice and record' (Whewell 1833a, xx; Yeo 1986a, 264–71). In effect, this meant that until a workable theory was agreed by the elite, it was difficult to instruct provincial cultivators in what to observe, although Whewell may have had some sympathy with Herschel's idea of a printed list of questions (Herschel 1830, 134; also 1849, iii).

Whewell's ruling on the relationship between theorists and collectors legitimated the mathematically adept Cambridge academics. As Morrell and Thackray suggest, it certainly relegated non-mathematical provincial amateurs to a secondary role in any of the advanced physical sciences (Morrell and Thackray 1981, 270–1). But Whewell did not mean that the theorists of all sciences were those engaged in the elaboration of mathematical laws. In fact, in one of his reports he advised that there were situations in which such elaboration could be inappropriate – for example, in sciences such as mineralogy and geology, where assiduous collection of data and its classification were priorities (Whewell 1832b). Furthermore, as discussed earlier, the chapter in the Bridgewater treatise, which Morrell and Thackray cite as evidence of his celebration of mathematics, was largely a critique of the moral effects of extreme deductive and mathematical habits of mind in science. His main point about the division of scientific labour was that theory and observation were intimately linked, not that deductive mathematics was superior to factual evidence.[5]

Whewell's criticism of the Baconian division of labour involved a

---

[5] In the case of electricity and magnetism Whewell warned about excessive mathematical treatments which wandered in 'that deep and charmed labyrinth much longer and further than the demands of physical science required' (Whewell 1835b, 29).

large investment of authority in theoretical experts. It was also connected with an affirmation of the role of significant individuals, and when leading men of science discussed this their language resembled that of the Tractarian critics, with its opposition of individual and institutional interests. Thus in 1835 William Rowan Hamilton asked Whewell whether there was likely to be an outbreak of 'fierce democracy': that is, whether 'the huge mass of members shall continue contented with their comparative obscurity, while a few persons necessarily occupy the chief attention' (Hamilton to Whewell, 6 April 1835, WP, Add. MS. a. 205[107]). Even before the first meeting, Herschel told Whewell of his strong objection to some of the essential premises of the Association, insisting that the aims of influencing governments, compiling annual reports, and facilitating the advance of science, could all be better accomplished by individuals than by announcements 'coming ex cathedra from a body'. At worst, Harcourt's proposal could produce a *'democratic tyranny'* or a 'mob' in which the individual scientist would be trammelled: 'Perfect spontaneous freedom of thought is the essence of scientific progress' (Herschel to Whewell, 20 September 1831, in Morrell and Thackray 1984, 66–8). When James Forbes, Whewell's disciple in Edinburgh, addressed this issue at the meeting there in 1834, Herschel's views had clearly affected him. Forbes was enthusiastic about the Association but endorsed the kind of distinctions made by Whewell and Herschel. Thus he acknowledged that very few can 'devote themselves unreservedly to those great enterprises which require the *whole man*'; however, a division of labour was as practical in 'intellectual as in mechanical science' provided the activity was directed by 'one designing mind' (J. D. Forbes 1834, xxii, my italics; compare Hamilton, in Graves 1882–5, I, 315). The degree of investment in the moral and intellectual character of such theoretical leaders is suggested by the controversy surrounding the respective claims of Newton and Flamsteed on data related to the movements of the moon. This can be seen as a test case for Whewell's claims about the moral character of great discoverers.

CONFLICTING CHARACTERS: FLAMSTEED VERSUS NEWTON

In 1833 Francis Baily chanced upon some lost papers of John Flamsteed, the first Astronomer Royal. He published these in 1835 as *An account of the Reverend John Flamsteed*. The papers had been in the possession of Mr Abraham Sharp, and Baily explained that they were 'found some years ago in a box deposited in an old lumber room, filled with books and papers, which had been considered as of so little use that they were frequently taken out by the servants to light the fire' (Baily, 1833, 462 and 1835, xii–xv; also Galloway 1836, 359–60). Sharp had been Flamsteed's assistant and apparently became the custodian of some 120 letters, Flamsteed's unpublished autobiography, and various papers relating to the controversy surrounding the publication of his *Historia Caelestis*. The most explosive part of this archive was the correspondence between Flamsteed and Newton beginning in 1704 and extending to 1710, when Flamsteed was disturbed by 'another piece of Sir Isaac Newton's ingenuity' (Baily 1835, xliii; cited in Baily 1836, 89). This referred to the fact that certain members of the Royal Society had been appointed by the Queen, on Newton's recommendation, as visitors to the Observatory. This group proceeded to publish Flamsteed's observations without his authority and corrections. Among the 175 sheets were records of 'the places of the moon' which he had given in confidence to Newton in 1694 (see Westfall 1980, 541–50, 655–66 for an account of this dispute).

In presenting this material Baily interpreted it as throwing new light on Flamsteed's character, a corrective to the unfavourable portrayals in other accounts, such as those by Lord North and Brewster (Baily 1835, xvi–xvii). He did not dilute the embarrassing evidence about Newton's behaviour in this dispute, although he suggested that he was 'rather the dupe of Halley's intrigues, than the original mover in so unworthy a contest' (1835, xxii). But the issue was inflamed when J. H. Barrow, writing in the *Quarterly Review*, presented Baily's new material as a blow to Newton's unsullied reputation (Barrow 1835, 96, 108). It was here that Whewell joined the debate, having been already urged to do so by Stephen Rigaud, the Professor of Astronomy at Oxford. In Rigaud's opinion the stakes were very high; he told Whewell that 'if Newton's character is lowered, the character of England is lowered, and the cause of Religion is injured' (Rigaud to Whewell, 25 January 1836, WP, Add. MS. a. 211[79]). In fact, Whewell was probably already aware of the damaging contents in the recovered correspondence, because Baily had asked George Peacock to inquire whether Whewell knew of any manuscripts at Trinity that might 'set the character of Newton, in this business, in a fairer point of view'. Baily did so because an earlier conversation had led him to believe that Whewell was intending to write a 'life of Newton, or something connected therewith' (Baily to Peacock, 11 January 1834, WP, Add. MS. a. 200[206]). Whewell had written, or at least drafted a letter, to Brewster answering his inquiries about Newton, saying that he was pleased a life of Newton was being written, and telling him about sources, including the manuscripts of Abraham Pryme and the Portsmouth collection (Whewell to Brewster, 2 October 1828, WP, R 18. 9[14], pp. 13–14).

Whewell's *Newton and Flamsteed* of 1835 was a reply to the article in the *Quarterly Review*. He acknowledged that the new documents had to be assessed by those 'who undertake the office of directing the judgement of the public', a role that he now seemed prepared to assume. In this case the issue was 'the good name of great men', and the *Quarterly* reviewer was asking readers to 'cast away all their reverence for the most revered name of our nation' (Whewell 1836, 3–4). Other commentators agreed that this was the implication of Baily's work; and in what Whewell regarded as a more temperate review Thomas Galloway explained the shock of this new material concerning Newton:

Succeeding writers viewed him at a distance, and the admiration universally bestowed on his discoveries in mathematics and philosophy was easily transferred to his private character. In the splendour of his intellectual greatness, the ordinary infirmities incident to humanity were unnoticed. (Galloway 1836, 392)

Galloway used the occasion to call for the publication of the Newton papers held by the Portsmouth family so that the character of Newton, 'a species of national property', could be fully assessed.

Whewell's reply indicates that what was at stake was not simply Newton's reputation as a man of science, but the status of the theorist in relation to that of the observer. While much of the reaction to the papers concerned the embarrassing fact of Newton's role in the Royal Society's seizure and publication of Flamsteed's work, Whewell gave equal focus to the 1694 dispute over access to the lunar observations. Here he constructed a contrast between Newton the supreme theorist and Flamsteed the humble astronomical observer who failed to appreciate the intensity of Newton's quest to bring the movements of the moon under the dominion of 'universal Gravitation' (Whewell 1836, 5). In seeking these observations Newton told Flamsteed that they would be worth far more if they were published in connection with the theory of gravity, for this would constitute 'a demonstration of their exactness, and make you readily acknowledged the greatest observer that has hitherto appeared in the world'. On the other hand, if they were announced without a theory to recommend them, they would be consigned to the 'heap of the observations of former astronomers'. Whewell unreservedly supported Newton's evaluation and explained Flamsteed's recalcitrance by alleging that 'he never fully accepted Newton's theory, nor comprehended its nature' (cited in 1836, 6–7).

In Whewell's opinion, Flamsteed, like other 'mere practical astronomers', conceived of theory as 'only a mode of expressing *laws of phenomena*, not a new generalization by which such laws are referred to a physical *cause*'. Thus Flamsteed spoke of his own theory of the moon, meaning projections on the basis of cumulative observations: '*I call it mine*, because it consists of my solar and lunar tables corrected by myself, and shall own nothing of Mr. Newton's labour till he fairly owns what he has had from the Observatory'. Whewell told the editor of the *Quarterly Review* that the distinction here was between 'the discovery of *what* occurred, and the discovery of *why* it occurred – between an observer and a philosopher' (Whewell 1836, 21–2). This was a distinction Flamsteed had failed to grasp.

When we note that Whewell's intervention in this controversy followed closely upon the reflections about the role of theorists in the British Association, it is possible to see its broader significance. The defence of Newton was not only the protection of a national hero, but an affirmation of 'the union of intellectual and moral excellence' – an assumption which several commentators saw as central to the debate (Whewell to editor of *Cambridge Chronicle*, in Whewell 1836, 28). Together with Herschel, Hamilton, and Forbes, Whewell believed that the assertion of the intellectual authority of theorists was partly dependent on their moral integrity since, as Forbes remarked, the highest activities of science required the 'whole man'. Whewell admitted that some of Newton's remarks seemed harsh but he related this to his enormous investment in the establishment of the theory of universal gravitation, to his lack of skill in managing disputes, and to the demands of his public office in the Queen's service. He supplemented this with a disparagement of Flamsteed's character: although conscientious and pious, the Astronomer Royal was 'suspicious, irritable and self-tormenting' (Whewell 1836, 9). The message of Whewell's account was that the observer in this dispute did not meet the intellectual and moral standards of the philosopher.

Thus Baily's *Account* of 1835 elicited a flurry of intense exchanges. Rigaud, who had persuaded Whewell to respond, saw the controversy as one of 'Flamsteedians' arraigned against the 'Newtonian confederacy' (Rigaud to Whewell, 25 February 1836 and 24 June 1836, WP, Add. MS, a. 211[82, 84]). Baily professed astonishment at the reaction and wrote a *Supplement* in 1837 which canvassed the major themes and, in doing so, opposed the stance of both Rigaud and Whewell. Demonstrating an impressive grasp of the evidence, Baily argued that its reception had been confused by the complexity of the controversy. For example, the two different disputes – over the access to lunar observations in 1693–4, and the subsequent one in 1710–11 over the publication of Flamsteed's catalogue by Halley at the direction of the Queen – were regularly conflated. But, most importantly, he suggested that misconceptions stemmed from 'an inattention to the precise state of the science at the period under review' (Baily 1837, 678).

The central point here concerned Flamsteed's understanding of Newton's theory – something that Whewell seriously questioned. Baily acknowledged that his interpretation was a direct rebuttal of 'some persons, whose talents I highly respect, and from whose judgement I seldom consider it safe to differ' (Baily 1837, 676). He meant Whewell. Baily started with Whewell's claim that the theory of gravity was not immediately established, and that 'no man of Newton's standing thoroughly accepted his views'. But he then used this point against Whewell's position: if Newton's theory was not well understood in 1694 then Flamsteed could hardly be accused of being ignorant of 'theory' if he did not accept it (1837, 679). Furthermore, Baily interrogated the use of 'the indefinite term "*Newton's lunar theory*"', an expression employed in the recent debate without adequate consideration of the precise meaning it carried '*at the period now under review*' (1837, 689). Indeed, his historical sensitivity went even further, for he claimed that the term 'theory' in Flamsteed's time normally denoted 'rules or formulae' for constructing astronomical diagrams and tables. This was an empirical or '*tabular*' theory, distinct from a '*physical*' theory providing a causal explanation of the behaviour of the moon. Baily argued that Newton's lunar theory at the time of the dispute with Flamsteed was not the fully 'physical' theory it later became after the mathematical labours of his successors, such as Euler, D'Alembert, and Clairaut. Each of these men was at first unconvinced by the theory because it did not 'fully accord with astronomical observations' (Baily 1837, 690).

The conclusion of this historical argument was that if difficulties remained into the eighteenth century, the so-called '*Newtonian lunar theory*' announced by Gregory in the scholium of the second edition of the *Principia*, fell far short of what is '*now* termed and understood as Newton's theory of the moon'. The earlier version was in fact a development of Horrox's rules for constructing lunar tables. In this sense Baily contended, against Whewell, that there could be no doubt that Flamsteed understood '*this portion*, at least, of what has been called the new lunar theory' (1837, 691–2). When Flamsteed reacted strongly to Halley's and Gregory's boast that Newton now needed no further lunar observations since the place of the moon could be deduced from theory, Baily contended that this indignation was justified. First, Flamsteed was right to complain that Newton had breached an undertaking not to circulate his theory before telling him; second, when he judged this 'theory' of Newton's in terms of its accordance with observations, he was following the scientific conventions of contemporary astronomy. In addition to this, if Flamsteed did not 'cordially *assent* to Newton's general theory ... it should not be considered as a reproach at that day, since there were many men, of much higher mathematical attainments, that had considerable doubts on the subject' (Baily 1837, 699).

Baily's detailed analysis was the only substantial challenge to Whewell's adjudication of this dispute. On this issue, at least, Baily did what no other contemporary could do: he contradicted Whewell's historical and philosophical authority – and survived. Whewell did not publish a rejoinder. Moreover, Baily explicitly rejected the strategy pursued by Whewell and Rigaud: defending Newton by 'attempting to lower the moral and scientific character of Flamsteed himself in public opinion' (Baily 1837, 676). There was a clear example of this in an article, probably by Rigaud, in the *Edinburgh Philosophical Magazine*, where Flamsteed was questioned by citing his 'belief in judicial astrology [and] ... his journey to Ireland to be touched for the benefit of his diseased body by a gentlemen, "whose gift", he tells us, "was of God"' (Rigaud 1836, 146). Baily's reassertion of Flamsteed's stature in the science of his day disrupted the move from intellectual to moral inferiority – a key part of Whewell's disquisition on the relative merits of theorists and observers (Baily 1837, 692, 699–700).

ALTERNATIVE VISIONS OF SCIENCE

The debate over the relative merits of Newton and Flamsteed exposed some of the difficulties involved in defining a model of appropriate scientific character. It shows quite clearly that, in spite of the growing importance of scientific institutions, there was a definite concern with the moral and intellectual qualities of the individual man of science, and especially of those figures who represented the high theoretical role. The responses to the quarrel between Newton and Flamsteed indicate that assessments of moral character informed arguments about the style of research and the relationships between theory and observation, between private and public knowledge. But this emphasis on character did not mean that there was a consensus about appropriate exemplars.

At the most general level there were significant tensions between the two major traditions available for the explication of the values and identity of the man of science. The tradition with the most ancient pedigree depended on a set of associations between the man of knowledge – the savant or natural philosopher – and solitude. The intersection of this notion with the ethos of modern science has been delineated by Dorinda Outram in her work on eighteenth-century French science, and more recently in Steven Shapin's discussion of seventeenth-century English experimental philosophy. In both cases, but with significant local inflections, there was a presumed link between the detachment of the savant or philosopher from the distractions of social and political life and his ability to reveal truths about the natural world. The scientific genius was thus often presented as a solitary individual.

In contrast with this, the influential view of the scientific enterprise propounded by Bacon in the early seventeenth century stressed the importance of the scientific community as the proper locus of the individual inquirer. Indeed, Bacon believed that the solitary individual would fall prey to what he called the idols of the cave – those mental idiosyncrasies peculiar to each individual. These had to be corrected by social intercourse. Thomas Sprat, in adulating the work of the Royal Society and its Baconian programme, described the Cartesian method as inferior because it was unsociable. Sprat's reading of Bacon was that science pursued in solitude was unreliable and even dangerous because it was unchecked by the scientific community (Shapin 1990a, 201–2). Yet, the full implication of

Bacon's view was that the institutionalization of science was no absolute guarantee against other idols of the mind, such as those that stemmed from the exchange of words and terms in scientific debate – the idols of the marketplace (Yeo 1985). Social discourse about science had its own problems and, for Bacon, these could only be overcome by the discipline of a method, one that made the talent and disposition of individuals largely irrelevant. It is worth considering the currency of these traditions in early-nineteenth-century Britain, since they represented the alternatives through which commentators such as Whewell sought to define the values of science and the proper behaviour of scientists.

Outram has shown how the pastoral ideal of the savant studying Nature in solitude, and achieving truth by means of this separation from the distractions of society, was deployed by Cuvier in his *éloges* at the Institut and later at the Académie Française. This allowed him to affirm a source of intellectual authority for the man of science independent of the fluctuating centres of political power (Outram 1978 and 1984, 91–2). Given her analysis of the ways in which this notion was deployed in specific scientific controversies and in relation to political struggles in France, it is understandable that it was not simply duplicated in Britain. Indeed, there is some evidence of explicit rejection. In reviewing the presidential discourses to the Royal Society in 1827, Brougham was critical of anything that emulated the 'French practice of *éloges*', precisely because they allowed a potentially divisive concentration on personality (Brougham 1827b, 353). This is not to say that the appeal to the notion of solitude as guarantor of knowledge was no longer available. The ideal of a secluded place for the interaction between the philosopher and nature was cultivated by Robert Boyle and other founders of the experimental programme in natural philosophy. It was therefore a tradition well established in British scientific culture. Indeed, it was expressed in the writings of Stewart, the custodian of the Scottish Enlightenment tradition which nurtured Brougham. In his *Philosophy of the active and moral powers*, Stewart spoke of the mind's 'natural direction' being distorted by the artificiality of commercial life (Stewart 1828, ii, 61, 85).

By the early nineteenth century, however, the ideal of solitude was hardly ever advanced in Britain without serious qualification. Whereas Cuvier could deploy this notion as a way of presenting science as a neutral and stable arena free from the chaos of political and social intrigue, leading members of British scientific institutions acknowledged its deficiencies. In his address as Secretary of the British Association in 1834, Forbes admitted that 'there has been . . . a general but most erroneous impression abroad, that philosophers are incapable of enjoying, and stoically superior to, the ordinary sociabilities of life, – that scientific ardour dwells only in the mind of the solitary'. His reply was that the meetings of the new body demonstrated the cooperation of many individuals, in a manner that broke down various solitudes: that of the country, or the 'still greater intellectual solitude of some noisy and commercial city' (Forbes 1834, xxi–xxii).

This theme was present in Hamilton's Dublin lectures on astronomy of 1832. He acknowledged that astronomical mathematicians may seem 'in the silence of their closets to have abandoned human affairs, and to live abstracted and apart'. However, in this lecture and in his address to the British Association in 1835, Hamilton insisted that 'genius in the very solitude of its meditations is yet essentially sympathetic' (Graves 1882–5, I, 650–55; II, 152). This may well have reflected the views of his mentor, William Wordsworth, who argued that poetic genius is not aloof, but nurtured by social 'sympathy'. Thus Hamilton claimed that discovery in science depended upon individual genius, but that such men were also 'influenced by the *Social Spirit*', a set of complex feelings cultivated by the comradeship of the scientific community (Graves 1882–5, II, 151–3). With slightly different emphases, both Forbes and Hamilton adjusted the notion of solitude to the Baconian rhetoric of cooperation and community. Although, like Whewell, they wanted to affirm the importance of theorists, this point could not be linked with an ideal of solitude without implying a criticism of the value of scientific institutions. It is significant that although Cuvier was quoted by Babbage as the model of a State-supported scientist, and by Lyell as the authority against transmutationism, his own rhetorical appeals to the ideal of solitude were not activated in the British debates (Outram 1976, 103–4).

It is likely that the idea of solitude was appropriated and radically transformed in ways that precluded its use by the scientific community. The comparison with France is again instructive because solitude was exploited in some of the critiques of the Académie des Sciences, before its closure in 1793. In the writings of Bernardin St Pierre, later taken up by the Jacobins, rural retreat and communion with nature were recommended as the means of insight into the natural world – an access denied to experts in the societies of the metropolis (Outram, 1984, 119; Hahn 1971, 138–9, 153). This notion of the untutored genius emerging from rural solitude to castigate the experts was not reactivated to any significant extent in Britain, not even in the popular educational rhetoric of the 1820s where it might have had some resonance. But in any case, by this time the idea of *withdrawal* from society as the basis for knowledge was largely overwhelmed by the Romantic conception of the creative individual aggressively *opposed* to society and its dominant values. These qualities were celebrated as heroic by the leading writers of the Romantic movement in Britain, and this made it even more difficult for men of science to endorse any notion of solitude.

REVIEW

This chapter opened with the claim that a concern with intellectual and moral character was an important element in early-nineteenth-century discussions of the nature of science. We saw that Whewell regarded inductive and deductive mental habits as both requiring and producing distinctive moral attitudes, thus making it difficult to advocate men of science *per se* as clear moral exemplars. Some of the debates generated at the British Association were informed by the issue of individual character, while at the same time conservative critics of scientific institutions complained that intellectual and moral direction, properly inhering in individuals, was being usurped by this new scientific body. Within the Association itself, this anxiety was partly shared by those who sought to protect the integrity and independence of elite theorists. This tension between different modes of science was personalized by the controversy over the Flamsteed papers, which coincided with these early debates. Whewell and Baily read the current arguments over the roles of theorists and observers into the dispute between Newton and Flamsteed in a manner that made these individuals embody levels in a hierarchy of scientific and moral activity.

Commentators such as Whewell who sought to affirm the authority of a theoretical elite had to ensure that its members were not identified with the ideal of the savant speculating in solitude. As noted above, this tradition was carefully qualified in Britain so that the role of theorists did not clash with the values of cooperation and association which were central to the Baconian ethos. In the hands of skilled apologists such as Herschel, the Baconian legacy became a flexible resource capable of legitimating the role of both observers and theorists (Yeo 1985 and 1986a). Indeed, in some ways Baconianism offered a framework in which the tensions between the individual pursuit of science and the wider social and institutional aspects of the endeavour could be resolved. For example, it avoided the problem of the *character* of the man of science, by stressing the role of methodology as distinct from personality. However, as we shall see, some of the leading scientific commentators believed that Bacon's methodology failed to provide an adequate account of significant discoveries by great men of science – precisely because it neglected the personal element.

NEWTON AS AN EXEMPLAR?

Just when the ideal of solitude had been largely absorbed and neutralized within Baconianism, the scholarly Victorian debate on Newton reopened the question of the appropriate ethos for the scientific community. Brewster's short *Life of Sir Isaac Newton* of 1831 combined an adulation of Newton as priest of science with a vehement rejection of Bacon's method and eighteenth-century convictions about Newton's debt to the Baconian programme. But this severing of Newton from the Baconian tradition had the effect of confirming the image of him as the solitary figure lauded by some of the Romantics. Furthermore, archival research did nothing to dispel earlier anecdotes about his eccentric habits and demeanour. Even before Brewster had full access to the Portsmouth papers, Whewell knew about Newton's alleged insane period. The focus on Newton threatened to revive the confrontation between alternative conceptions of the scientific ethos, because there was a danger that he could be represented as a scientific version of the solitary, asocial Romantic hero.

What were the implications of romantic notions of genius for the image of the scientist? There were two levels at which this operated. First, the Romantics promoted a concept of creative genius as transcending rules and conventions in poetry, art, and science. Second, they encouraged the idea that such genius was intimately bound up with an extraordinary personality – one capable of breaking with conventional methods to achieve great discoveries, but also likely to transgress traditional norms of behaviour (for more

details, see Yeo 1988; Knight 1967, 71). It is here, inevitably, that the
question of Newton's character appeared, because although his vision
of the world was condemned as mechanical by Blake and Coleridge,
another view, put by Wordsworth and Carlyle, associated his heroic
triumphs with the concept of solitary genius. Perhaps the most
evocative image occurred in Thomas Carlyle's essay on the 'Signs of
the times' of 1829:

No Newton, by silent meditation, now discovers the system of the world from
the falling of an apple; but some quite other than Newton stands in his
Museum, his Scientific Institution, and behind whole batteries of retorts,
digesters and galvanic piles imperatively 'interrogates Nature', – who,
however, shows no haste to answer. (Carlyle 1829, 443)

Here the status of Newton and his mode of discovery was contrasted
directly with the values of organized, cooperative Baconian science.
This contrast did not appear so explicitly six years later in the
controversy surrounding the Flamsteed papers, even though the
dichotomy between theorist and observer was central. Whewell may
have been aware of this other set of connotations when he urged that
Newton must not be seen as a 'Hero of a philosophical romance or a
scientific mythology', but rather as a man conducting business of
State (Whewell 1836, 14). If this was his attempt to dissociate Newton
from Romantic notions of genius, it was completely ineffective. It is
worth noting that in 1896 Merz still circulated an image of Newton as
one 'who had to retire into solitude' in order to escape the Baconian
emphasis on practical application and thus lay 'the more permanent
foundations of the new research' (Merz 1896, 1, 95). But in any case,
Whewell's own critique of Baconian method, which Merz was
endorsing, exposed it as incapable of accounting for the imaginative
leaps of heroic scientific discoverers such as Newton.

    Brewster, De Morgan, and Whewell were the three major com-
mentators on Newton in the early Victorian period. They all wanted
to present Newton as the exemplar of scientific achievement, but their
study of his work and behaviour reactivated the tension between the
ethos of solitude and that of Baconianism. Whewell and Brewster
agreed – as they seldom did – that the Baconian methodology was
profoundly incapable of accounting for the scientific discoveries of
figures such as Kepler and Newton. Brewster complained that any
association between Newton and Bacon's rules of method could only
'tend to depose Newton from the high priesthood of nature'. In his

view, Bacon did not appreciate that great discoveries took place in a manner which was 'the very reverse of the method of induction'. 'The impatience of genius', he protested, 'spurns the restraints of mechanical rules, and never will submit to the plodding drudgery of inductive discipline' (Brewster 1831b, 330–7). As we shall see in the next chapter, Whewell took a similar line in his *History*, leading Brewster to charge that Whewell had taken the critique of Bacon from his earlier *Life of Newton* (for De Morgan's views, see De Morgan 1915, I, 75–84; Theerman 1985; Yeo 1988).

The problem with this critique of Baconian method was that it ran the risk of linking the recognition of imagination in science with the ideal of solitary genius celebrated by the Romantics. Carlyle clearly connected these two when he pictured Newton as achieving spectacular discoveries independent of scientific institutions. In his writing, this idea of isolation became one of active antagonism to various social conventions and, even apart from its anti-Baconianism, could not be tolerated by the scientific community. But the less aggressive ideal of solitude was also inadmissible because of its perceived association with dangerous personality and moral character. The case of Herschel is instructive. When seeking, in his *Discourse*, to promote science as an activity open to people of all social stations, he described it as 'a most delightful retreat from the agitations and dissensions of the world', thus drawing on the pastoral tradition discussed earlier (Herschel 1830, 16). But some of his friends later felt that a preference for solitude was a *personal* problem for him. Maria Edgeworth wrote to Harriet Butler in 1843 and spoke of Herschel's shy and sensitive character, suggesting that he 'is on the verge of the dreadful danger of over wrought intellect – over excited sensibility'. On the basis of other knowledge she offered this analysis:

In his earlier life when disappointed in love he shut himself up in darkness . . . and even of later times when vexed in friendship or when scientific things go wrong he betakes himself to darkness and solitude and in abstraction shuts himself up from the external universe. Very very dangerous! Solitary confinement even to the innocent – and even to the stupid a very hazardous experiment! (Edgeworth to Butler, 3 December 1843, in Edgeworth 1971, 596–7; Schweber 1981b, I, 37–8)[6]

How did Newton's admirers seek to detach him from this notion of solitary genius? Brewster did not attempt to, although he was perhaps

---

[6] Herschel defended Newton's moral character in his review of Whewell's works (Herschel 1841, 180–9).

the first biographer to reply to the contention of J.B. Biot that Newton had suffered emotional depression and mental derangement in 1693 (Biot 1821, 25–37; Yeo 1988, 273–6). In fact, Brewster almost cited the solitary mode of Newton's life as evidence *against* any problem of character:

The unbroken equanimity of Newton's mind, the purity of his moral character, his temperate and abstemious life, his ardent and unaffected piety, and the weakness of his imaginative powers, all indicated a mind which was not likely to be overset by any affliction to which it could be exposed. (Brewster 1831b, 224–5)[7]

De Morgan's approach questioned the mythology of inspirational discovery which had gathered around Newton's achievement. In 1837 he gave a satirical rendition of this view:

the Newton of the world at large sat down under a tree, saw an apple fall, and after an intense reverie, the length of which is not stated, got up, with the theory of gravitation well planned, if not fit to print.' (De Morgan 1837, 242–3)

In his opinion, this was a myth constructed by Newton himself, but it was also the product of a fault shared by other discoverers, who concealed errors, trials, and guesses, and presented their conclusions as the result of an elaborate train of deduction. This misrepresented the process of discovery, including that of Newton's, which was 'more like a book-keeping operation than the poetical process of the fable' (De Morgan 1837, 242–3). Furthermore, De Morgan argued that the image of Newton as a solitary figure, an untutored genius who could dispense at an early stage with the work of Euclid and Descartes, overlooked the manner in which his genius was shaped by Cambridge. 'The lad', he said, 'carried to the University as much conceit as the man brought away of learning and judgement' (De Morgan 1914, 8–12). In one respect, then, this was a response that drew on the institutional emphasis within Baconianism, although De Morgan noted that this could have negative as well as positive effects on science. While Brewster saw Newton's move to the Mint as one of the rare moments of state recognition of science, De Morgan did not hide his dismay: 'And where should a high-priest of science have lived and died? At the Mint? (De Morgan 1855, 337.)

Whewell reminded those who followed the controversy over the

---

[7] In his second, larger work on Newton, Brewster did suggest that Newton had moved from intellectual solitude to the 'society of men' at Cambridge (Brewster 1855, I, 20).

Flamsteed Papers that Newton's discovery was not made at once, and his account in the *History* can be interpreted as another counter to the idea of Newton as a solitary genius. Although in other places he did stress the Cambridge context, here he suggested that, however limited Newton's social contacts were, he was not solitary in an historical sense; rather, his genius, or sagacity, as he called it, was able to work on the ground prepared by others (Whewell 1836, 5; and ch. 6 below).

However, these remarks did not effectively address the question of Newton's moral character which, on the argument of the Bridgewater treatise, was so crucial in the case of such an outstanding inductive discoverer. At the same time as he was making claims about the moral effects of inductive and deductive mathematics in this work, Whewell was asserting the value of Newtonian synthetic geometry against Continental analysis in the undergraduate curriculum. Here, he sought to present Newton not just as a superlative practitioner of this method but as the embodiment of the virtues it represented – clarity of fundamental conceptions, rigorous exposition of argument, and striking inductive generalization (Whewell 1832e, vi, xx; also ch. 8 below). This was not an investment that could withstand the candidness of Powell's assessment, which accepted the troubled nature of Newton's personality:

The truth is that the intellect which had most deeply sounded and explored the mysteries of external nature was at times perplexed and obscured by the mysteries and infirmities of its own constitution, and in embracing the system of the universe Newton at times lost possession of himself. (Powell 1834a, 534)

In the *History*, Whewell reasserted the notion of Newton as exemplifying the fact that 'great talents are naturally associated with virtue', making no reference to Biot's allegations or to his own dispute with Baily. However, he did make the following oblique response to the issue of Newton's character which these previous debates had raised:

Often, lost in meditation, he knew not what he did, and his mind appeared to have quite forgotten its connexion with the body... Even with his transcendent powers, to do what he did, was almost irreconcileable with the common conditions of human life; and required the utmost devotion of thought, energy of effort, and steadiness of will, – the strongest character, as well as the highest endowments, which belong to man. (Whewell 1857a, II, 141)

Unwittingly, this account was an eloquent reflection of a transition from the view that genius is accompanied by virtue to the acknowledgement that it has to resist vice. It posed the question invited by Whewell's contention, in his Bridgewater treatise, that inductive discoverers were moral characters: namely, what followed if they were not? This was especially pertinent if discoveries were ascribed to the imaginative capacities of individuals, rather than to rules of method. In the absence of this moral foundation, was the scientific process deprived of some of its intellectual security? In the contemporary debates, assumptions about moral character were linked with arguments about styles of scientific behaviour: the authority of theorists in the British Association, for example, was conceived in moral as well as intellectual terms. Newton's relationship with Flamsteed was adjudicated within a similar rubric. But neither Whewell, Brewster, nor De Morgan was able to reinstate unequivocally the alliance between intellectual and moral virtue. The Tractarians would not accept Newton or any other secular figure as a moral model. The result was not only that Newton, the most celebrated man of science, remained a deeply ambiguous exemplar for the scientific community, but that the project of describing an appropriate scientific personality remained intrinsically contentious. Whewell had been caught in these difficulties, and when he published the first of his major works in 1837 he seemed to suggest that a moral could be discerned in the history of science rather than in the character of its greatest exponents.

CHAPTER 6

Using history

... at Lord Northampton's the other days one of the Charades
they acted was in honor of Whewell: – they began by
repres[enting] 'hewing' a tree, then a 'well': in hunting for truth
they find a book (Hist [*sic*] of Ind. Sciences), on wch Fame seizes
and afterwards delivers to immortality: – the Party then
crowned Whewell with laurel...

Romilly's Cambridge Diary, 1 January 1838, Bury 1967, 137

When Whewell received a medal from the Royal Society in 1837 for
his researches on tides, the president, the Duke of Sussex, seemed
more captivated by the recent appearance of *History of the inductive
sciences*. In making the award the Duke referred to Whewell's 'last and
highest vocation' – that of the historian who had traced 'the causes
which have advanced or checked the progress of the inductive
sciences from the first dawn of philosophy in Greece to their mature
development in the nineteenth century' (Todhunter 1876, I, 87–8).
But Whewell was less sure of the reception of this work. 'Of my
History of science', he remarked in 1840, 'the principal notice taken
by men of science has been of a hostile kind; and I do not think that
any practical cultivators of special sciences will feel any deference for
a person who has presumed to speculate about them all' (Whewell to
Lord Northampton, 5 October 1840, in Todhunter 1876, II, 293).
When declining an invitation (which he eventually accepted) to serve
as President of the British Association, he pleaded that 'My *History*
and *Philosophy* of Science are disqualifications, not qualifications, for
being put at the head of the scientific world' (Whewell to Murchison,
18 September 1840, in Todhunter 1876, II, 286).

HISTORY AS METASCIENCE

These reflections relate to the peculiar position as metascientist that
Whewell had cast for himself. As we have seen, this role began to

emerge in his major reviews during the early 1830s; but the scale of the *History* was regarded as a major bid on Whewell's part for the judicial role conventionally accredited to historians. Writing to Jones in 1836 Whewell was candid about his ambitions, saying that he had 'pleasant fancies of doing something in the way of philosophizing which shall have a permanent place in the history of the world' (Whewell to Jones, 4 July 1836, WP, Add. MS. c. 51[199]). Lyell acknowledged the value of this synthetic function, telling Whewell that 'being a Universalist' rather than a specialist, he was helping the advance of science (Todhunter 1876, I, 112).

Whewell's two major publications, the *History* and the *Philosophy*, are closely related. In July 1834 he described them as parts of a three-stage project. The third part, on the moral sciences, was never completed, but the first two parts – the major works published in 1837 and 1840 – were written over the same period. The first was intended to be a study of the inductive sciences, 'historiographized in a new and philosophical manner'; the second was more 'metaphysical and transcendental'. Whewell said that when this more philosophical mode took over, 'I open *Book two*' (Whewell to Jones, 27 July 1834, in Todhunter 1876, I, 90; and 6 October 1834, in II, 193). Research on the two projects overlapped, and in the prefaces to both works Whewell stressed their interdependence and their common concern with the *historical* development of the sciences – 'the most certain and stable portions of knowledge which we already possess' (Whewell 1847a, I, 1). Thus in discussing Whewell's historical perspective in this chapter it will sometimes be necessary to refer to the *Philosophy* and to other writings after 1837, because his views on the history of the sciences were expressed in various texts and depended on the context in which they were expressed.

In his recent work on Whewell's philosophy of science, Fisch argues that the *History* was premature, in that it was completed without the benefit of the fully developed epistemology presented in the *Philosophy*. At a fine level of philosophical detail, such as the degree to which it tolerated 'alternative theorizing' or the extent to which it fully conveyed the 'creative role of the mind', this difference is noticeable (Fisch 1991a, 115–16). For example, ideas are sometimes described as being 'applied' to 'special and certain Facts', rather than as constituting these facts (Whewell 1857a, I, 7, 181). But in terms of Whewell's performance as a major public commentator on science, it is doubtful that the book suffered from this incomplete philosophy.

Indeed, Fisch's point about the lack of fit between the two works, in spite of Whewell's original plan, suggests the need to take the *History* on its own terms (see also Cantor 1991a, 74). After all, it still contained significantly unconventional views on most of the topics usually covered in similar works.

When speaking of the *Philosophy* Whewell said he was writing 'not for scientific so much as for general, or at least metaphysical readers' (Whewell to Lubbock, 7 July 1840, in Todhunter 1876, II, 284). Presumably it was also this audience, larger than the scientific community, to which the less philosophically technical *History* was also addressed. Whewell saw this book as one that might interest a wider readership, and when he wrote to his sister in 1837 he reported that 'it has sold remarkably well', despite Brewster's negative review (Whewell to Martha, 18 November 1837, in Stair-Douglas 1881, 189). In dedicating it to Herschel he mentioned the popularity of his friend's *Discourse*, and hoped that his own book might 'have a chance of exciting an interest in some of your readers' (Whewell 1857a, I, v).[1]

The decision to write a history meant both opportunities and problems for Whewell's metascientific commentary. On the positive side it offered a greater scope for generalizations about the nature of science than the review journals. In these, the reviewer had to evaluate a particular text for a specialist group, while also trying to capture the general reader. In contrast, the historical form had already been used to present lessons about the advance of science, such as the importance of method or experiment. The negative side was that Whewell's ambitious attempt to embrace all natural sciences, including quite recent disciplines such as geology and physiology, ran the risk of failing to satisfy specialists, while also complicating the general story sought by another audience. For the more expert readers, the book possibly defined those scientific subjects outside their immediate province. Thus George Peacock praised the mineralogy, geology, botany, mechanics, and chemistry, and said 'the metaphysical parts' were excellent. But he regarded the treatment of astronomy, closest to his interests, as 'the least successful' (Romilly's Cambridge Diary, 21 May 1837, in Bury 1967, 119). This is why Whewell was uncertain about the reception of his efforts in the increasingly specialized world of science (Whewell 1837a, I, xiii).

The *History* was, as Elkana and Cantor have stressed, a novel and

---

[1] The *History* (but not the *Philosophy*) was translated into German by J. J. von Littrow in 1840.

grand achievement (Elkana 1984; Cantor 1991). But before considering its message and significance we need to ask why history of science was important. Put simply, it was because history was the master narrative of the period. In the year in which Whewell's work appeared, Thomas Carlyle produced his *French Revolution*. When Mill reviewed this book, the first draft of which had been earlier burnt by his housekeeper, he called it 'an epic poem' (Lepenies 1988, 103). This was precisely the high moral status accorded to the genre by a variety of contemporaries: for example, those influenced by the tradition of philosophical history in Scotland, those entranced by Walter Scott's historical novels, those inspired by the quite different work of Barthold Niebuhr in Germany, or by English writers such as Carlyle and Macaulay who claimed that historical writing was an imaginative art. Historical narrative was also seen as a basis from which judgements – moral, political, and philosophical – could be made (Collini, Winch, and Burrow 1983, ch. 6). This is what Whewell meant when he referred to the 'judicial position' involved in the 'functions of the historian'. Brewster did not agree with the judgements Whewell made in this role, but he acknowledged the medium of history as one through which persons such as Whewell and Comte could legitimately seek to become 'the legislators of science, the adjudicators of its honours, and the arbiters of its destiny' (Brewster 1838a, 273).

In his *Discourse*, Herschel suggested that such an account was urgently needed. Explaining that such a 'regular analysis of the history of each [scientific] department' was beyond the scope of his own book, he hinted at Whewell's plans: 'We are not, however, without a hope that this great desideratum in science will, ere long, be supplied from another quarter every way calculated to do it justice' (Herschel 1830, 219–20). Herschel of course realized that there were quite substantial historical surveys, such as those in the preliminary dissertations to the *Supplement* of the *Encyclopaedia Britannica* from 1815, but his hope for something more was shared by others. In reviewing the first three reports of the British Association, De Morgan referred to the 'description of the actual state of the various branches of philosophy, which cannot be properly done without a complete knowledge of their past history' (De Morgan 1835, 156; also Powell 1837; Galloway 1844, 198). This was essential, as Herschel said, for any knowledge of the 'general principles' that guide the 'progress of discovery' (Herschel 1830, 219). Given the existing role of historical accounts and the call for more detailed ones, Whewell could not avoid this genre if he wished to pursue his metascientific vocation. But what assumptions and approaches were available to Whewell as he embarked on the *History*?

HISTORICAL CONSCIOUSNESS IN THE EARLY VICTORIAN PERIOD

Several scholars have described the period from about 1800 to 1830 in terms of 'the rise of historical consciousness'. This perhaps begs more questions than it answers; but the case for a major development in historiography rests on two important grounds: first, that there was a new critical use of sources, especially primary, archival documents; and second, that there was a greater sensitivity to the social and cultural differences between various historical periods. These early decades of the nineteenth century also saw the emergence of new disciplines such as geology, archaeology, anthropology, and the study of language, all having strong historical frameworks (Levine 1986).

With reference to these developments, A. N. L. Munby remarked that 'there is nothing mysterious about the sudden growth of the study of early science. Indeed, it would have been odd if it had escaped unstudied' (Munby 1968, 2). He noted that from the 1820s there was a strong interest among men of science in the earlier periods of scientific activity, particularly in the revolutionary achievements of the seventeenth century and the great names associated with them. There was even a short-lived Historical Society of Science, founded in 1841 by the twenty-year-old James Orchard Halliwell, who had just left Cambridge after one year as an undergraduate (Dickinson 1932; Hornberger 1949). Powell and De Morgan, but not Whewell, were members of its Council, which listed a set of works for printing. Although the Society folded within the year, the programme of systematic research into the surviving primary records was already established, and by the middle of the century there were important biographies of Galileo, Kepler, and Newton, scholarly editions of Bacon's works, and collections of letters and papers on seventeenth-century science (see Rigaud 1838 and 1851). Potentially at least, discussions about the scientific past could now be more finely textured than those of the previous century. In 1779 David Hume had said that because modern astronomy was accepted, it was now only 'a matter of mere curiosity to study the first writers on that subject'

(Hume 1948, 24). In contrast, by the early nineteenth century, statements on past science were significant elements in contemporary debates about both particular theories and the nature of the scientific enterprise. This meant that while there were efforts to catalogue the primary documents, there were also strong pressures to use them in the service of present commitments (Graham, Lepenies, and Weingart 1983).

When Whewell began his *History* in the early 1830s there were two main approaches to the history of science: Enlightenment surveys of scientific progress and histories of particular disciplines. Whewell's work drew on each of these, but in ways that challenged their dominant assumptions.

Enlightenment surveys

The main rationale for histories of science was the belief that a record of past discoveries could reveal the method by which science progressed. General surveys of science, usually dealing with the more established branches of the mathematical and physical sciences, often made this point. The *Supplement* to the fourth, fifth, and sixth editions of the *Encyclopaedia Britannica*, published between 1815 and 1824, included preliminary dissertations that offered historical surveys of science. The editor, Macvey Napier, described them as attempts at tracing 'the *history* of Philosophy and Science' (Prospectus, July 1815, NLS, 1948. 63[50]). John Playfair and John Leslie both wrote on the physical and mathematical sciences, and Thomas Brande on chemistry. These works shared two key messages: that true science had a difficult birth and that the chances of error were still high, and that from the time of Bacon and Newton a successful method was available and its continued application would produce further progress. Playfair underlined the lessons involved: 'compared with natural philosophy, as it now exists, the ancient physics are rude and imperfect... Science was not merely stationary, but often retrograde.' Yet early scientific ideas were worth considering because they 'elucidate the history of the errors and illusions to which the human mind is subject' (Playfair 1822, II, 67).

This is what has come to be called a 'Whig' perspective on the history of science (Butterfield 1931; Oldroyd 1980). Past science was largely approached with present theories as criteria for evaluating earlier thinkers and theories. Second, the past was usually divided into two discrete periods – one of failure and one of success – with the passage from darkness to light being ascribed to the adoption of the correct method by celebrated heroes. Finally, significant points of transition such as that associated with the influence of correct method, or the appearance of a major theory, were sometimes referred to as 'revolutions'; the subsequent development of the science being conceived as a steady accumulation of empirical data under the compass of a general theory. In this way, histories of science incorporated a discourse about continuity and discontinuity that was central to contemporary political historiography.

Histories of disciplines

The most common vehicles for historical accounts of science were books, encyclopaedia entries, or addresses dealing with particular disciplines. The crystallization of specialized subject areas and their institutionalization in separate scientific societies created the opportunity, and the need, for narratives that located the genealogy of present activities. By the first decade of the nineteenth century the preference for histories of single disciplines, or areas, rather than surveys of general science was already established. Thus in his *History of the Royal Society* of 1812, Thomas Thomson explained that 'to give the reader a greater interest in the sciences ... it was thought necessary to begin the history of *every* science as nearly as possible at its origin, and to give a rapid sketch of its progress' (Thomson 1812, preface; my italics).[2] Textbooks on particular branches of science, encyclopaedias, and *éloges* on the achievements of departed scientists all deployed historical accounts in order to highlight the state of present knowledge. Given the significant, and often radical developments in several areas of science from the 1790s, these prefaces or chapters did not simply chronicle a list of discoveries, but sought to make a case for certain theories and methods associated with a particular conception of the subject. While there was a concern to

---

[2] The bibliographical survey of histories of science in R. M. Gascoigne (1984, 993–1,019) suggests that most examples between 1753 and 1837 concentrated on particular disciplines: only six titles (apart from Whewell's) dealt with two or more sciences, and these usually did not range beyond the mathematico-physical sciences. More commonly, histories of science focused on a particular discipline, or a set of related subjects. Whewell said he borrowed from both 'the histories of the special sciences and of philosophy in general', although Brewster reprimanded him for inadequate reference to other works (Brewster 1837, 112–13). The forthcoming bibliography of historical works on science by Rachel Laudan will allow comparative observations.

anchor each discipline in the past there was also a tendency to relegate some parts to a pre-history from which the new science had escaped. At one extreme, Xavier Bichat, an advocate of physiology as a new, independent domain, regarded the past as largely irrelevant: 'Our era no longer needs historical monuments. The history of science must be put to one side; it is science itself which we must present' (cited in Outram 1986, 352). But history was too valuable as a polemical resource for Bichat's indifference to take hold. More usually, defenders of new disciplines or approaches provided a history that showed the errors of the past in order to illustrate the pitfalls from which the favoured doctrine, or hero, had rescued scientific knowledge.

How did Whewell's *History* compare with these existing approaches? Apart from the range of sciences he included, Whewell went entirely against the dominant British tradition by using history to advance a non-Baconian account of science. Thus when Julius and Augustus Hare described the *History* in their *Guesses at truth* of 1848 as the work of Bacon's 'most enlightened disciple', they may not have appreciated its full implications (Preyer 1981, 65). In fact, Whewell was consciously seeking to show that the historical record revealed a more complex and interesting picture of science than Bacon, or at least his most ardent admirers, including Herschel, had supposed. By exploiting motifs such as progress, revolution, and tradition – common in contemporary political history – Whewell presented the development of science as a drama containing the conflicts and moral issues found in other human affairs. In using these themes he also contested some of the connotations that they had in Enlightenment historiography. Thus his idea of progress did not include technology, commercialism, or industrialism; tradition was not synonymous with obscurantism; and revolution did not always imply a complete break with the past, and was certainly not a warrant for radical political change.

This admittedly limited survey provides a set of themes through which Whewell's *History* can be discussed: first, the Whig notion of progress – in this case the gradual, cumulative progress of scientific knowledge towards nineteenth-century doctrines: second, the idea of historical study as a means of revealing the true scientific method; and third, the concept of revolution in science and its connection with Whig political historiography.

PROGRESS

The precise genesis of Whewell's *History* is difficult to trace, partly because his notebooks are not consistently dated; but it is fairly clear that it developed in conjunction with his study of induction. His notebooks on this date from 1830, but even earlier he wrote to Jones, saying: 'I still meditate doing something about the History of the Metaphysics of Mechanics though as yet it is only an intention' (Whewell to Jones 23 September 1822, WP, Add. MS. c. 51[15]). In his early reviews, it was the historical success of the natural sciences that caught Whewell's attention as he tried to define and affirm inductive philosophy. He attacked the tendency of certain writers, especially political economists, to propound deductive principles as dogmas when they were not adequately grounded in observation. As an alternative, the success and progress of physical science offered a 'new model and a new hope', and he believed that the method which had led to this knowledge must be fruitful when applied to 'moral and political as well as physical subjects' (Whewell, unpublished Notebooks, WP, R 18. 17[6 and 8]; for the commencement of his induction project and the historical inquiry, see R 18. 17[2 and 8]).

Not long after these initial sketches, Whewell published *First principles of mechanics, with historical and practical illustrations* of 1832, one of the series of textbooks dating from 1819 (Fisch 1991a, 51, 60–1). Unlike political economy, mechanics was a mature science and could be presented as a series of deductions from well-established principles. But in this text Whewell argued that mechanics should be taught as an inductive science which had progressed by confronting a series of empirical problems (Whewell 1832c, iv–v). During the following year, he appears to have confronted the possibility of extending this approach to other sciences. In a notebook entitled 'The philosophy of the progressive sciences', he projected a study of the 'past history of the most complete branches of human knowledge', making it clear that it was this 'state of completeness' which would allow 'important lessons concerning the nature of knowledge, the constitution of man's mind, [and] the act of discovering truth'. These remarks, probably made in late 1833, represent the seeds of the *History*; between 1832 and 1837 Whewell widened this historical account beyond the most perfect sciences of astronomy, mechanics, and optics to include the growth of the less complete or infant sciences (Whewell, Notebooks,

WP, R 18. 17^8, p. 1; see also pp. 1–13; the last entry is dated 29 December 1833).

There are certainly parallels between Whewell's plans and the conclusions of earlier histories of science. Like Playfair, he saw history as a means of revealing the method of past and future progress in science. But there is evidence of Whewell's early dissension from some of the assumptions of this prevailing historiography (WP, R 18. 17^8, pp. 5–6). In his diary, for example, an entry dated 26 September 1827 lists the 'Vulgar errors of the nineteenth century'. These included the idea of the 'superiority of the age', 'Utility (vs feeling and imagination)' and the notion that 'science now precedes by induction by Bacon's rules' (Diary, WP, R 18. 9^{14}). By the time of his address to the British Association in 1833 Whewell found it strange that some critics of modern science should see in it a source of overweening pride in the 'superiority of the present generation, and of the intellectual power and progress of man'. Only in astronomy had men been able to 'complete' their knowledge; in optics a similar success seemed assured. But the remaining prospect was 'comparatively darkness and chaos; limited rules, imperfectly known'. Significantly, Whewell suggested that these feelings of pride probably derived from the 'natural exultation which men feel at witnessing the successes of art', and he stressed the need to distinguish between knowledge and its application, between theory and practice, between science and art (Whewell 1833a, xxiii–xxiv). These demarcations were the premises of the *History*, allowing Whewell to omit the practical and mechanical arts (technology) unless a major scientific principle was involved. In doing this, he severed all links with some of the popular hymns to science, contending that its legitimation was not utilitarian, but moral and intellectual (Whewell 1857a, 1, 255; see ch. 8 below).

These examples suggest that Whewell was critical of the current historiography of science before he developed a definite anti-empiricist epistemology. One possible context for this departure is his work on the history of architecture. In 1830, following two European tours, Whewell published *Architectural notes on German Churches*, in which, as the subtitle suggested, he speculated about the origin of the Gothic style (Becher 1991, 4–7). Although apparently more distant from his scientific commentary than the textbooks on mechanics, this task forced him to confront the case of a transition between cultural systems. As Whewell put it, the introduction of the 'pointed arch' was part of the process in which 'the old system of architecture, derived from the classical styles, was finally converted into one of a different and opposite kind'. Significantly, he rejected the prevailing notion that '*Gothic* was synonymous with *barbarous*', and sought an understanding of the principles 'which govern its forms' (Whewell 1830a, iii, xvi–xvii). Thus before he began writing history of science, Whewell was sensitive to the possibility of appreciating the integrity of an intellectual phenomenon, rather than simply judging its deficiency against a prevailing convention. Seven years before the *History*, the following passage indicates a distinctive approach to the past:

The features and details of the later architecture were brought out more and more completely, in proportion as the *idea*, or internal principle of unity and harmony in the newer works, became clear and single, like that which pervaded the buildings of antiquity. (Whewell 1830a, iv)

Here the term 'idea' carries some of the weight it later assumed in Whewell's account of science. In the second edition of 1835 he spoke of the need to understand Gothic style as 'a connected and organic whole' (cited in Whewell 1842, 2–3). With this awareness of the intellectual integrity of former architectural styles, it is not surprising that he was dissatisfied with existing accounts of past scientific theories.

It is also important to recognize that Whewell had links with other Trinity scholars such as Julius and Augustus Hare and Connop Thirlwall who, with Thomas Arnold, were influenced by the German historian, Barthold Niebuhr. These 'Liberal Anglican' historians, as Duncan Forbes has called them, strove to emulate the new standards of German scholarship, stressing the importance of primary documents and the particularity of time and place. Their work questioned the assumptions of eighteenth-century rationalism and its optimism about the inevitable progress of reason. When applied to the history of thought, including science, this perspective cautioned against a dismissive attitude to the past and regarded present progress as dependent upon past achievements (D. Forbes 1952; Cantor 1991, 75–8).

One of Whewell's major innovations was the use of a three-stage model of scientific progress (Cantor 1991, 71). Taking as a starting-point the case of astronomy – his 'pattern' science – he sought to show that scientific fields or disciplines move through a 'prelude', an 'inductive epoch', and a 'sequel'. Whewell explained that it was this perspective, and not the simple narrative of facts, which made his work distinctive; and looking back at the time of the third edition in 1857 he believed that this scheme had been largely accepted as 'fairly exhibiting the progress of scientific truth' (Whewell 1857a, I, vii). Terms such as epoch and sequel, and associated ones such as 'stationary period' and 'revolution', were not new. These appear in earlier histories by Playfair, Brande, and Thomson; but Whewell linked them together in a general historical theory that was intended to be a basis for the 'Philosophy of the sciences' (Whewell 1840a, I, 4, 8–11 for the novelty of his approach; for references to stationary periods and epochs, see Playfair 1822, II, 58, 67, 441).

Whewell's writing was rich in metaphor – geographical, architectural, political – but the overriding structure of the *History* was theatrical. Herschel recognized this, remarking that Whewell's novel approach gave 'a picturesque or rather epic interest to his narrative' by highlighting the most important epochs in the history of science (Herschel 1841, 187). These were the 'Inductive Epochs', the turning-points of the drama he unfolded; they were the moments when the inductive process acted in a 'more energetic and powerful manner', when distinct 'Facts' and clear 'Ideas', the two key elements of scientific knowledge, were brought together by the mind of a great discoverer and a new truth was seized and fixed forever. However, Whewell claimed that the excitement of such epochs was not unprepared, and that closer study revealed a 'Prelude' in which the relevant facts and ideas had been 'gradually evolved into clearness and connexion' (Whewell 1857a, I, 5, 7, 10; see Whewell to Jones, 5 August 1834, WP, Add. MS. c. 51^{174} for an early expression of this theme). After the inductive epoch associated with the names of one or two principal discoverers there followed a 'Sequel' in which the discovery was accepted and extended by other men of science. In this way Whewell combined competing images of scientific advance: the discontinuity of the inductive epoch with its distinctive breakthrough, and the continuity of slow preparation and consolidation in the prelude and sequel. Later in this chapter we will see that a similar combination, or tension, characterized contemporary political histories. For the moment, we can return to the issue of Whewell's attitude to the concept of progress.

In introducing the *History*, Whewell said that the '*progress* of knowledge' was the 'main action of our drama', and there is no doubt that the emergence of general theories successfully explaining empirical phenomena was a strong feature of this work (Whewell 1857a, I, 4). There is a definite similarity between Whewell's partisanship on behalf of the wave theory of light in his treatment of optics, and the Whiggish histories of disciplines such as geology (Cantor 1983, 1–3). But in contrast with most of these histories, Whewell's vision was not one of continual progress away from past obscurantism; rather, it suggested long 'stationary periods' (Whewell 1857a, I, 11–12). The history of science was certainly not a manifestation of improving mental capacities, or the 'March of Mind', as the satirist Thomas Peacock dubbed the self-confidence of his contemporaries. And unlike eighteenth-century writers such as Priestley and Condorcet, who celebrated recent achievements as emancipations from past errors, Whewell made a case for the serious study of failures in science, arguing that they disclosed important clues about scientific discovery. This was partly because his *History* was not a mere record of discoveries but an analysis of the dialectic between facts and theories, sensations and ideas – later elaborated in his *Philosophy* as the 'fundamental antithesis' of all knowledge (Whewell 1847a, II, 647–68; see ch. 1 above). Since a lack of balance or fit between facts and ideas was the major reason for scientific error, modern science was also potentially susceptible to failures (Whewell 1857a, I, 5–7).

Whewell's historiography did not endorse a simple contrast between modern science and past obscurantism. Confident about the progress of science, he was interested in the reasons for past failures and did not dismiss erroneous doctrines as foolish. As Elkana has noted, when Whewell discussed failures in science, such as those of the Greeks, he spoke of the ideas involved as 'often vague, indeed, but not, therefore, unmeaning' (Whewell 1857a, I, 27; Elkana 1984, xviii). In writing of the mistakes and achievements of past science, Whewell, unlike the rationalist historians of the Enlightenment, displayed some sensitivity to cultural relativism, recognizing that social factors were relevant to the kind of science practised in particular times and places, just as Jones had shown in the case of forms of rent. This meant that Whewell's claim that Western Europe was the centre of science was historical, rather than simply chauvinistic. Admittedly, though, he did confess that he would not be upset if his reading showed that 'Alhazen the Arabian' had discovered nothing in optics, 'for I have a strong *European nationality*' (Whewell to Jones, October 1837 in Todhunter, 1876, II, 261; Cantor 1991a, 78).

After the advances of the Greeks, the Middle Ages were a period of quiescence in the history of science. If these centuries were not entirely dark, they were certainly 'stationary'. Later historians seeking to defend the achievements of this period had to contend with Whewell's picture of its thinkers as failing to carry on the advances of the Greeks in mechanics and astronomy (Whewell 1857a, 1, 183–91).[3] His explanation was that they failed to cultivate clear 'fundamental scientific ideas', and this in turn was linked with the authority of the Church. The thinkers of this age were also possessed by the spirit of mysticism, which pursued either particulars or wild generalizations. Thus 'their physical science became Magic, their Astronomy became Astrology, the study of the Composition of bodies became Alchemy' (Whewell 1857a, 1, 215, 233). Another negative factor was 'the commentatorial spirit' – the disposition to 'read nature through books'. This later hardened into the verbalism and dogmatism of the Scholastics, for whom Whewell reserved his harshest judgement (1, 200, 203, 237–51).

Nevertheless, Whewell still differed from the majority of writers who used the failures of past science as a means of glorifying modern successes. For example, it was common for histories of chemistry to begin with a lurid account of alchemy in order to congratulate the founders of chemistry on liberating the subject from such irrationalism. But Whewell subtly undercut this convention by rejecting any simple connection between medieval alchemy and modern chemistry: there was no substantive intellectual link. By refusing to allow alchemy as the 'mother of Chemistry', Whewell deprived the disciplinary histories of their story of emancipation. Conversely, he suggested that as one factor in the revolt against scholastic authority, alchemy may have assisted the rise of scientific thought (Whewell 1857a, 1, 232–3, 246).[4]

In their historical writings both Lyell and Powell castigated the baneful influence of past thought and described a movement from a period of intellectual darkness to the illumination of modern doctrines. Of course, there were strategic interests behind this approach. Thus Lyell sought to tell a story in which the new science of

---

[3] In 1860 Jacob Burckhardt continued this negative image of the Middle Ages in *The civilization of the Renaissance in Italy*, but this was later challenged by Pierre Duhem, who argued that there was a continuous line of scientific development from the medieval schools. See Rosen 1964, 80–1.

[4] For a conventional account, see T. Thomson 1830, 1; Macaulay later repeated the essence of this approach in his *History of England*. See Macaulay (1913–15), 1, 400–2.

geology, together with the human mind, was liberated from the superstitious, anthropomorphic, and theological prejudices of the past (Porter 1976). Powell's similar account of the effect of past beliefs was less colourful, but it was part of his effort to clarify the boundaries between science and theology. It allowed him to present teleological thinking as a vestige of an earlier, less scientific stage. Powell was attracted to Comte's idea of three stages of intellectual development – the theological, metaphysical and positive, or scientific stage – and viewed the development of the various sciences as a movement 'all in one direction', away from metaphysical notions to 'the recognition of regulated causes, law and order' (Powell 1855, 63–74 and 1839).

In contrast, Whewell stressed the dependence of scientific progress on the debates and speculations of the past, including those that Lyell and Powell regarded as unscientific. Certainly, he did say that the 'commentatorial spirit' of the stationary period impeded scientific progress because it encouraged exegesis rather than experiment: 'criticism took the place of induction; and instead of great discoverers we had learned men' (Whewell 1857a, I, 204). But like mysticism, this 'speculative tendency' manifested a concern with ideas, and stimulated abstract thought by which 'scientific Ideas' were distinguished from 'common Notions'. Thus the mysticism of Kepler inspired his scientific thinking; although Whewell was quick to add that this effect was possible because Kepler possessed clear scientific ideas (Whewell 1857a, I, 12–13, 319–20). Whewell stressed that such speculation could not be divorced from observation and experiment – as it had been during the Middle Ages – but equally insisted on the necessity of the discussions that Comte condemned as 'metaphysical'. Continuing this point in the *Philosophy*, he wrote that 'physical discoverers have differed from barren speculators, not by having no metaphysics in their heads, but by having good metaphysics' (Whewell 1847a, I, x). This had not been recognized by those writers 'accustomed to talk with contempt of all past controversies, and to wonder at the blindness of those who did not at first take the view, which was established at last'. These metaphysical controversies brought various subjects 'into a condition in which errour is almost out of our reach' (1847a, II, 377–8).

Having examined the ways in which Whewell qualified the conventional statements of progress, it is important not to forget that this concept, in the sense of a wider coverage of empirical facts by laws, was central to his work. Like the other histories, Whewell's narrative was directed towards the present. In fact, he maintained that 'we do not learn the just value and the right place of imperfect attempts and partial advances in science, except by seeing to what they lead' (Whewell 1857a, III, 135). Once a successful theory had emerged it was possible to appreciate its connections with previous doctrines. But such links were more complex than the image of a liberation from the past obscurantism. This meant that Whewell could not accept the views on method usually associated with histories of scientific progress.

METHOD

It is fair to say that Whewell believed he knew what induction was before he began a detailed attempt to illustrate it in the history of science. In the early 1830s he foreshadowed a study of various branches of science that described how each commenced its path to inductive generalization – from observation to laws to causes (Whewell, Notebook, WP, R 18. 17[5 and 15]; p. 46 of the latter is dated 2 July 1831). This compares with the account of scientific method in his review of Herschel. In the *History*, Whewell proposed that a survey of past discoveries 'may not only remind us of what we have, but may teach us how to improve and increase our store', and 'afford us some indication of the most promising mode of directing our future efforts to add to its extent and completeness' (Whewell 1857a, I, 4). It was thus conceived as a part of a neo-Baconian project for classifying the sciences, their development, and the means of future progress.

But in the *History* Whewell abandoned the orthodox Baconian account of induction assumed in earlier notebooks, and defended the speculative guesses of Kepler as the more usual mode of great discoveries. His study of the progress of science provided the material for his theory of method, subsequently expounded in the *Philosophy*. In both works, Whewell used the term induction to refer to the general process by which laws and theories were attained; but he stressed that this was more than a mere generalization from the facts, because it involved the addition of a conception from the mind of the scientist. Although he stressed that 'facts and ideas' were involved in the 'formation of science' in the introduction to his *History*, he made this philosophical point more directly in *The mechanical Euclid*, also published in 1837:

Some notion is *superinduced* upon the observed facts. In each inductive process, there is some general idea introduced, which is given, not by the phenomena, but by the mind. (Whewell 1837c, 178; also 187; Fisch 1991a, 101)

These ideal conceptions, such as that of the ellipse in Kepler's astronomy, derived from the fundamental ideas – for example, space, number, resemblance – appropriate to a particular branch of science. Whewell claimed that these, and other ideas, 'regulate the active operations of our mind', and are the grounds of the necessary truths which certain branches of science had so far established (Whewell 1847a, I, 66). Furthermore, the ability to derive from a fundamental Idea the appropriate conception, capable of distilling a collection of facts into a law, required the 'peculiar Sagacity which belongs to the genius of a Discoverer' (1847a, II, 40). Although the epistemology underlying this position was not fully elaborated until the *Philosophy*, the *History* showed its methodological implications. These were at odds with current assumptions about the possibility of public involvement in science.

During the Enlightenment, writers such as Joseph Priestley regarded the history of science as a collection of lessons and clues for those seeking new discoveries. In 1767, for example, he argued that the various branches of science would be assisted by 'an historical account of their rise, progress and present state', since such an inquiry 'cannot but animate us in our attempts to advance still further, and suggest methods and experiments to assist us in our further progress' (Priestley 1775, I, vi). This confidence was sustained by a belief that discoveries of electrical phenomena had been made more by 'accident' than by 'human genius'; and extending this into a more general programme, Priestley called for the compilation of histories for each branch of science and the encouragement of wider participation. This was feasible because 'many modest and ingenious persons may be engaged to attempt philosophical investigations when they see, that it requires no more sagacity to find new truths, than they themselves are masters of' (Priestley 1775, II, 166; also I, xviii; and see McEvoy and McGuire 1975, 325–404). The lessons of Priestley's historiography were egalitarian: a closer examination of great discoveries diminished their heroic quality and suggested that they were achieved by methods open to all.

In the 1830s, the legacy of Enlightenment assumptions about

knowledge was visible in the works produced by the Library of Useful Knowledge. Some of the scientific biographies in this collection continued the strategy of reducing genius to method or accessible procedures.[5] Thus the volume on Galileo presented him as an exponent of the safe method of experiment and induction recommended by Bacon. But from this perspective the success of Kepler was embarrassing, because it had to be admitted that 'this extraordinary man pursued, almost invariably, the hypothetical method':

His life was passed in speculating on the results of a few principles assumed by him, from very precarious analogies ... We nevertheless find that he did, in spite of this unphilosophical method, arrive at discoveries which have served as guides to some of the most valuable truths of modern science. (Drinkwater 1833, i)

In attempting to salvage a moral from this case the writer indicated that Kepler always abandoned hypotheses if they did not fit with facts; but this was seen as small compensation for his persistent efforts in 'seizing truths across the wildest and most absurd theories' (Drinkwater 1833, 2, 15).

In the previous chapter we saw how Whewell began, at the British Association meeting of 1833, to criticize the notion of unguided collection of data. Four years later, in the *History*, he legitimated the role of imagination in the major theoretical advances of science. Significantly, he referred to this biography of Kepler, remarking that several persons:

seem to have been alarmed at the *Moral* that their readers might draw, from the tale of a Quest of Knowledge, in which the Hero, though fantastical and self-willed, and violating in his conduct, as they conceived, all right rule and sound philosophy, is rewarded with the most signal triumphs. (Whewell 1857a, 1, 317)

Whewell then used the case of Kepler to assert that the process that led to great scientific discoveries involved imagination and speculation, not the 'cautious or rigorous process' endorsed by some followers of Bacon (Whewell 1857a, 1, 318; for the wider debate, see Yeo 1985). He dealt with Newton's much-quoted remarks about industry and patient thought as the key to his discoveries by suggesting that this was a fair account of the 'mental *effort*' involved in science, but

---

[5] In his best-selling *Self help*, Smiles announced that 'in the pursuit of even the highest branches of human inquiry, the commoner qualities are found the most useful'; genius was 'common sense intensified' (Smiles 1894, 94–5, 317).

distinguished this from 'the natural *powers* of men's minds [which] are not on that account the less different' (Whewell 1857a, II, 140).

Like most other writers, Whewell saw history as a source of lessons for the best means of making further discoveries. But rather than viewing these as likely to derive from the application of a method open to all, the *History* attributed major discoveries to the uncommon characteristics, both intellectual and moral, of exceptional individuals: Copernicus, Kepler, Newton, Lavoisier, Fresnel, Faraday. The inductive epochs – those periods in which significant inductive leaps were made in various fields – centred on key discoveries or generalizations of a 'fundamental conception' in the work of particular scientists (Whewell 1857a, I, 9–10). And although the *History* aimed at learning something of the method by which progress in science occurred, Whewell concluded that truly great men of science, in their epoch-making contributions, went beyond any specifiable rules of method. While he was prepared to describe the necessary characteristics of a great scientific mind – 'distinctness of intuition, tenacity and facility in tracing logical connection, fertility of invention, and a strong tendency to generalization' – this was hardly the basis for an egalitarian method of discovery. Whewell did note the unsuccessful guesses of discoverers, but not with the intention of endorsing democratic notions, such as those of Priestley, about the nature of discovery (1857a, I, 317–31; II, 139 and 1847a, vii–viii).

Whereas most contemporary histories of science regarded scientific discoveries as the result of proper method, Whewell gave new significance to individual qualities transcending method. Herschel certainly saw it in these terms when he approved this break with the eighteenth-century Enlightenment attitudes carried on by the utilitarians:

It is too much the present fashion to ascribe all progress – at least all modern progress – in inductive science ... to 'the Age', as if there were some magic in the word, and as if by its use it were possible to elude or abate down the acknowledgement of individual pre-eminence. (Herschel 1841, 187)

Whewell agreed with Herschel and made a similar point in one of his textbooks in 1832, contending that the notion of great discoverers as 'products of their age' gave inadequate conceptions of figures such as Newton (Whewell 1832e, x). But given the discussion in the previous chapter about the problem of demonstrating the combination of intellectual and moral virtue in these individuals, it is significant that the *History* set their work firmly within its intellectual context. Thus while Newton's achievement required the special qualities mentioned earlier, it was not simply an individual achievement but an historical one – 'the catastrophe of the philosophic drama' (Whewell 1837a, II, 139). Whewell analysed it into five separate steps indicating how, in bringing them all under one theory, Newton was able to take as '*facts*' the '*laws*' established by his predecessors (1837a, II, 117, 136–8).

While Whewell stressed the importance of scientific genius and its peculiar achievements, he combined this with the aim of establishing a 'philosophy of science'. In so far as this meant a set of generalizations about the nature of scientific progress, there was always a degree of ambiguity in the *History*. This was one reason for Brewster's unfavourable comments in the *Edinburgh Review*. There were several levels to Brewster's assault, many of them activated by his perception of Whewell as the head of a Cambridge coterie hostile to all Scotsmen (Morrell 1984, 29). One crucial point involved a dispute concerning the meaning of the term 'art of discovery'. Whewell denied that there was any simple, mechanical set of rules for producing major advances in science, warning that there was no 'Popular Road' to truth (Whewell 1847a, II, 366; for the tensions in Whewell's views on method, see Yeo 1986a, 270–1). Brewster agreed with this and endorsed Whewell's reading of the case of Kepler, even claiming that the anti-Baconian lesson drawn by Whewell was derived from his own earlier *Life of Newton* of 1831. But Brewster argued that Whewell was indeed seeking to 'give laws to inventive and original minds', to 'heaven-born genius', by promoting vacuous generalizations about the relationship between facts and appropriate ideas (Brewster 1837, 117–25, 150; also De Morgan 1840, 709). For Brewster, no significant scientific discoveries could be reduced to rules because the relevant factors, including accidental discovery (a category disallowed by Whewell), were unpredictable. It is also significant that when he attacked Baconian method in his work on Newton, Brewster claimed that if 'the general character' of scientific discovery was to be found, it would not be through history, but in 'the biographies of eminent men' (Brewster 1837, 126 and 1831b, 332–6).

Whewell was obviously annoyed by Brewster's attack and even wrote a mock review of the *History* in which he gave it favourable treatment and answered some of the criticisms (WP, R 18. 10[5]). Whewell's public response was that in spite of the element of inventive genius in the achievements of figures such as Kepler, Fresnel, and, indeed, Brewster, there was still value in an attempt to 'analyse and methodize the process of discovery' (Whewell 1840a, II, 186–7 and 1837e; Brewster 1842a, 288–92). He also distinguished between the capacities involved in foundational discoveries and those required for scientific work after the intellectual framework of the discipline was established. In this respect the *History* sought a delicate balance between the dramatic inductive leaps by great discoverers and the more routine scientific work of the sequel that followed each inductive epoch. Whewell regarded this time as crucial, because it determined whether the work of 'great geniuses' would expire with them or be consolidated by a body of followers. 'This', he said, 'is always a work of time and labour, often of difficulty and conflict' (Whewell 1857a, I, 10). In Herschel's interpretation of this notion the period of the sequel resembled 'the occupation and settling of the country under the dominion of the conquerers, quelling the insurrectionary movements of ignorance and prejudice under the new regime, and partitioning out the land in provinces and domains' (Herschel 1841, 186; Yeo 1987, 32). In this respect, the *History* not only concerned the unpredictable imagination of men of genius, but more codifiable factors leading to the establishment of an inductive theory within the scientific community. It may be possible to say that Whewell was not advancing rules or methods for making epochal discoveries, but was seeking to generalize the pattern by which these discoveries were translated into scientific disciplines (Schaffer 1986, 387–420).

REVOLUTION

I. B. Cohen has noted that the phrase 'revolution in science' occurs in Whewell's comments on Copernicus, Bacon, Harvey, Locke, and Lavoisier. But it is difficult to endorse Cohen's contention that Whewell's use of the term 'revolution' was part of a 'fully developed theory', unless we recognize the tensions surrounding this concept in his work (Cohen 1985, 529–32). If we compare the two previous sections there is a tension between Whewell's stress on the dependency of scientific advances on past contributions, and his account of the creative element in major inductive discoveries. Given this, what did 'revolution' mean for Whewell?

Although the term 'revolution' occurs in connection with discussions of scientific change in the *History*, there are several issues raised by Whewell's usage. First, there is the question of when revolutions occur. The most obvious answer is in the inductive epoch when the great theoretical leaps are made. But in the case of physical astronomy, Whewell stressed the degree to which major breaks were made with past doctrines during the prelude to Newton's work, leaving him to crown this 'metamorphosis' (Whewell 1857a, II, 113–14, 116, 137). Second, there is the question of their frequency. Here it is necessary to recall that Whewell was dealing with the history of sciences, not simply science as single entity; thus he could refer to revolutions in astronomy, optics, and chemistry.

Third, it is necessary to consider the way in which Whewell spoke of revolution in connection with the method or philosophy of science. Here Bacon was the hero who had reinforced the movement against scholastic attitudes, announcing a 'New Method' and turning 'the Insurrection into a Revolution' (Whewell 1857a, II, 40–1). Recently Frits Schipper has claimed that this is the least ambiguous way in which Whewell uses the term 'revolution'. He suggests that Whewell distinguished between those changes in scientific theory that were accompanied by a revolution in method and those which were not (Schipper 1988, 44–5). Thus the revolution in physical astronomy at the time of Newton was the more dramatic because it combined with a revolution in method – the one whose spirit, if not letter, was expressed by Bacon. In contrast, the 'revolution' of Lavoisier was accomplished more smoothly because it could rely on the previous methodological changes, or, as Whewell put it, the 'great improvement' in the means of arriving at truth. Hence Lavoisier, unlike others who had 'produced revolutions in science' saw his theory accepted within a very short time (Whewell 1857a, III, 119–20).

Another problem here is that it is not clear whether Whewell believed that the revolution in method was completed, once and for all, at the time of Bacon and Newton. In Book XI of the *Philosophy*, he traced the continuing controversy over the relative emphasis given to '*Ideas* and *Sensations*' in theories of knowledge. He regarded his own work as an attempt to achieve the right balance (Whewell 1840a, II, 283). Finally, there is the difficult issue concerning the character of 'revolutions' in Whewell's work. This arises because there are apparently contradictory statements in the *History*. For example, in the introduction he spoke of 'revolutions of the intellectual world' as being 'steps of generalization', thus giving the impression of gradual change – inconsistent with the notion of a dramatic shift that Cohen discusses.

Having sketched the different senses of 'revolution' in Whewell's writing, it is useful to place them in the wider context of general historical writing. Two observations recommend this: first, Whewell drew heavily on political history for his metaphors, and second, contemporary historiography also revealed tensions between notions of revolution and gradual change.

Whewell's use of the term 'revolution' requires some scrutiny, since the political connotations of the word in the 1830s were far from favourable. British politics of this decade were conducted against the background of the July 1830 overthrow of the Bourbon monarchy in France, and of course, against the terrible symbol of the French Revolution itself. Even the compromises surrounding the Reform Bill of 1832 met with strong conservative opposition, and Whewell's known Tory leanings – at least by the time he was appointed Master by Peel – suggest a less-than-enthusiastic attitude to hasty political and social change. But some accounts of Whewell's conservatism are exaggerated. Wordsworth, who strongly opposed the Reform Bill, assumed that Whewell had sympathized with the petition to admit Dissenters because he had supported John Lubbock, a radical candidate at the Cambridge elections in 1832 (Wordsworth to Whewell, 14 May [1834] in Hill 1978–9, v, 710). Five years earlier, Whewell had discussed the Catholic question with him and argued that although change for its own sake should be resisted, 'human institutions' must 'accommodate themselves to the ever changing relations and forms of society' (Whewell to Wordsworth, 20 January 1829, in Hill 1978–88, 48–9; Schaffer 1991, 205). But Wordsworth had no great cause for concern. As Master, Whewell's very delicate revision of the statutes governing the college produced no radical change, effectively leaving Trinity in 1846 under regulations dating from 1560 – a situation later addressed by the Royal Commission investigating the ancient universities (Robson 1967, 333).

In discussions with H. J. Rose in 1826 Whewell opposed the view that advances in science involved the rejection of 'what was known for something new' (see ch. 3 above). Later, in considering the possibility of arriving at socio-political knowledge, he told James Marshall that such a search must not begin with the notion that 'the conditions of this generation' were 'novel and unparalleled', but with the awareness that 'there is a great deal of truth already in the world, and a great deal of it embodied in the frame of society'. Change and renewal in both Church and State should proceed from 'a formative spirit which makes *reform* unnecessary'. When reflecting on his own work, he remarked that 'however much I may feel a craving for a new view of the truth, suited to our age, . . . I have no hope of finding any view of the truth, which is really true, if it do not include and rest upon that which has been true up to the present time' (Whewell to Marshall, 27 December 1842, in Stair-Douglas 1881, 281–3; see also Whewell 1834d, 21–22; and 1838, 132–3; Schaffer 1991, 206).

Thus in both the political and intellectual spheres, although Whewell believed in progress, he was sceptical about the notion of radical change involving drastic breaks with the past. In this sense he shared, to some degree, the dilemma of the Whig political historians: that of balancing an approval of change and reform with respect for values and traditions enshrined in the past.

In his study of Victorian historiography, J. W. Burrow has restored complexity and nuance to the image of Whig history. While accepting Butterfield's emphasis on the confidence and optimism of this outlook, Burrow draws attention to another dimension of Whig historians: their sensitivity towards the past. Even in the most ardent celebrations of present achievements, such as those in the writings of T. B. Macaulay, there was a tension between progress and continuity. Although Whig history was a success story, it was not one that ridiculed the past: the past was revered because it embodied the tradition by which progress was sustained (Burrow 1981, 2–3). In this respect, the Whig historians cannot easily be equated with the Enlightenment rationalists whom Forbes identified as the opponents of his Liberal Anglicans. Indeed, on Burrow's reading, the Whigs, like the Liberal Anglicans, may have imbibed the 'Burkean tradition', in so far as the notion of 'change in continuity' was central to their historical consciousness, one in which the stress on progress was constrained by the idealization of continuity (Burrow 1981, 22–3, 47–8, 52; for the influence of Edmund Burke on the Liberal Anglicans, see D. Forbes 1952, 1, 6).

This account opens the way to further readings of Whewell's *History*. Without denying the insights offered by the comparisons with Liberal Anglicanism, it is possible to take advantage of Burrow's portrait of Whig historiography. For example, the Whig acknowledgement and approval of progress – both intellectual and social – is closer to Whewell's outlook than the cyclical vision contemplated by the writers identified by Forbes as Liberal Anglicans. At the same time, the problem of reconciling a support for institutional adaptations and political reforms with a commitment to the continuity of a particular tradition, appears to have an analogue in Whewell's conception of scientific change. In both cases, the notion of revolution becomes problematic. Thus in Lord Acton's view, Burke and Macaulay regarded the settlement after the 'Glorious Revolution' of 1688 as conservative rather than revolutionary: 'it was little more than a rectification of recent error, and a return to ancient principles' (Cohen 1985, 71). Burrow draws attention to the fact that it was the Tories and radicals who could entertain images of dramatic ruptures in English history; whereas the Whigs' need for continuity compelled them to see bridges between past and present (Burrow 1981, 16–17, 34).

An awareness of these tensions in Whig historiography provides a perspective on Whewell's ambivalence about the concept of revolution. This is not to suggest that there is no sense of dramatic and significant change in the *History*. Whewell regarded the successful inductive generalization of an inductive epoch as a breakthrough that placed the relevant science on a new intellectual footing and set the scene for significant redirection, subsequently recorded by changes in terminology – a process he compared with 'the change of the current coin' accompanying 'great political revolutions' (Whewell 1857a, I, 9). As noted in the previous section, he celebrated the genius of individual scientists – Copernicus, Newton, Lavoisier – who grasped the ideal conceptions necessary for these discoveries and, in this sense, acknowledged their often revolutionary departure from preceding scientific ideas. In the case of Lavoisier he rejected the claim of those who argued that the French chemist did nothing substantially new, indicating that in this case and in others the originality of a great discoverer was 'proved by the conflict' which their work aroused in the scientific community (1857a, III, 119).

But at times Whewell drew back from the possible analogies between dramatic political and scientific change. Thus he did not make any positive link between the ideological upheavals of the civil war in seventeenth-century England and the intellectual advances of Newton, Boyle, and Hooke; in fact he suggested that scientific activity was an *escape* from the political turmoil (Whewell 1857a, II, 112). And in the revolutionary year of 1848, when discussing Frederick Myers's book on Great Men, Whewell criticized the view that the 'great man is above rules', telling Myers that 'it leads you at times to speak as if rebellion, insurrection, and revolution were good and great, in general and in the abstract' (Whewell to Myers, 18 June 1848, in Stair-Douglas 1881, 351).

It is not surprising, therefore, that Whewell restrained the possible implications of this account of scientific change in ways that underlined the continuities between past and present. Thus, although the crucial discovery of an inductive epoch required a grasp of the relationship between facts and ideas not achieved by the previous generation, Whewell noted the extent to which the 'principal discoverer' was preceded by 'trials, seekings, and guesses, on the part of others' (Whewell 1857a, II, 99). He reinforced this point when considering the dispositions of major scientists: 'Undoubtedly this deference for the great men of the past, joined with the talent of seizing the spirit of their methods when the letter of their theories is no longer tenable, is the true mental constitution of discoverers' (1857a, I, 286). Sciences were not formed by a single act but by a long series of progressive changes involving 'apparently contradictory' principles:

The principles which constituted the triumph of the preceding stages of the science, may appear to be subverted and ejected by the later discoveries, but in fact they are, (so far as they were true,) taken up into the subsequent doctrines and included in them. They continue to be an essential part of the science. The earlier truths are not expelled but absorbed, not contradicted but extended; and the history of each science, which may thus appear like a succession of revolutions, is, in reality, a series of developments. (Whewell 1857a, I, 8)

This perspective allowed the celebration of significant advances constituting an 'epoch' together with an acknowledgement of their dependence on past efforts. It is here that the idea of a conservation of past constitutional wisdom, apparent in the historical vision of Burke and Macaulay, has its analogue in Whewell's image of a 'vast patrimony of science' in which real achievements live on:

Thus the final form of each science contains the substance of each of its preceding modifications; and all that was at any antecedent period discovered and established, ministers to the ultimate development of its proper branch of knowledge. (Whewell 1857a, I, 3, 8; for a refinement of this view, see Whewell 1851b)

Like the Whig historians Whewell sought a compromise between notions of revolutionary change and gradual development, between progress and continuity. This was necessary because he wanted to acknowledge the major conceptual shifts in science while still stressing the significant links between new and older theories. But this tension could not be entirely resolved because it derived from Whewell's insistence on the novel and imaginative element of major inductive discoveries.

THE MORAL OF THE STORY

In his original preface Whewell explained that the aim was to deduce 'lessons from the past history of human knowledge', but that this would now be done in the *Philosophy* (Whewell 1837a, 1, 5). When Sedgwick suggested that the historical narrative needed a moral, Whewell said this would be as long as the story. But it might also be that the *History* did not deliver any simple lesson. Whewell certainly came to believe that it did not offer a popular one, remarking in the second edition that the sympathy of 'popular readers' goes with the path leading to 'false science and to failure' (in Whewell 1857a, 1, 339; also Whewell to Marshall, 19 August 1854, in Stair-Douglas 1881, 435–6).

This may have something to do with the way his work subverted, or at least complicated, some of the dominant historical themes associated with the concepts of progress, method, and revolution. Although these were important elements in the historical drama he unfolded, each of them represented a point of tension in the work, partly because in each case he qualified existing meanings. Like the Enlightenment surveys and the disciplinary histories, Whewell affirmed progress, but in doing so did not support the image of scientific advances as complete breaks with an obscurantist past. The history of science was not entirely one of successive inductive generalizations leading gradually towards modern doctrines; rather, older, erroneous theories were replaced by new theories, often in sudden and dramatic fashion. This observation related to his rejection of a strict Baconian method as the mode of major scientific discovery and his emphasis on the imaginative leaps of heroic figures. Yet at the same time he wanted to advance a philosophy, as opposed to an art, of discovery – a set of generalizations about the processes by which science advanced.

Given this, Whewell's concept of revolution was complex. He did not support the notion of a single, revolutionary liberation from the dark pre-history of a science: rather, he stressed the important intellectual preparation in periods before the first crucial inductive discoveries of a particular science. There was often more than one significant change within the history of an individual discipline. Thus, even in formal (positional) astronomy there were revolutionary shifts from Hipparchus to Copernicus to Kepler. The implication here was that more recent disciplines, immature by comparison with astronomy, could not assume that their present theories were immutable. In this way, the triumphalism of the disciplinary histories was tempered by the contrast with other branches of science in Whewell's more comparative account.

Whewell's statements on the nature of scientific change did not end with the *History*; nor did the ambiguities. In fact, his emphasis seemed to vary according to the opponents addressed. Thus when replying to the cumulative, empiricist view of scientific development – such as that used by Macaulay in a speech of 1846 on the triumphs of modern knowledge – Whewell stressed that knowledge did not usually grow by 'repeated *addition* of part to part, but by perpetual *transformations*' (Whewell 1849a, 173). In order to oppose the gradualist image of Macaulay's case, the term 'transformation' here carried connotations of profound, qualitative change. Two years later, however, Whewell used the notion of transformation as a counterweight to that of radical or revolutionary conceptual shifts in science. Here he gave a more positive estimate of Descartes than in the *History*, arguing against the view that Newtonianism represented a complete rupture with Cartesianism (Whewell 1851b, 493). This was partly connected with his attempt to defend Isaac Barrow and other Cambridge Fellows from the charge of being slow in rejecting Cartesianism.[6]

The attempt to combine a history of individual disciplines within a general historical survey of science was a distinctive feature of Whewell's work. In the mock review mentioned earlier, he began by saying that 'although we are ourselves no students of any of the other *ologies* in any technical shape, it must be possible to write such a history of science as shall be intelligible and interesting' (WP, R 18. 10[5]). Yet with continuing specialization it became impossible for one man to keep pace with the detail required for an historical analysis of the kind he attempted. Even at the time of the first edition he

---

[6] Whewell stressed that 'the mechanical and optical explanations of natural phenomena given by Descartes were the best which the world had then seen' (Whewell 1859, iv–vi), and said that English philosophers admired Descartes, 'even when they rejected his doctrines'. This compared unfavourably with 'the disparaging tone which prevailed from Newton's time – the latter a piece of gross servility' (Whewell to De Morgan, 9 February 1859, WP, O 15. 47[26]).

admitted to Richard Owen that the new field of physiology was outside his competence (Whewell to Owen, 30 November 1837, Cambridge Library Add. MS. 5354 E 8). By the third edition these problems were severe, and although Whewell told Forbes that he wished 'to give some indications of the progress made in science since the last edition was published', he also admitted that any additions would have to be 'very slight and general' (Whewell to Forbes, 18 July 1856 and 8 October 1856, in Todhunter 1876, II, 406–8). After reading George Cornewall Lewis's *History of Astronomy* he expressed awe at its detail. This suggests that he realized that a full rewriting of his own *History* was impossible (Whewell to Lewis, 29 January 1862, in Todhunter 1876, II, 424).[7]

Whewell's problem was not only one of keeping abreast with specialist fields; he also faced the task of making them tell a general story. Beginning with the 'pattern' science of astronomy, he attempted to analyse the development of other sciences, focussing on the stages of their development, the crystallization of clear ideas in connection with relevant facts, and the consolidation, in each case, of a scientific terminology. However, the historical record did not always supply the data to support his generalizations: only six out of seventeen disciplines fitted the three-stage model of growth, and most did not display the successful inductive laws of the mature sciences. Consequently, the inductive charts which, like 'the Map of a River', were intended to show the path by which particular facts had been combined to form general truths, had only two examples – astronomy and optics (Whewell 1857a, I, 11 and 1847, II, 118; see Schaffer 1991, 230 on the case of electrochemical sciences).

In reviewing the final edition of the *History*, Forbes acknowledged its value as a grand survey, saying that it would remain a 'permanent work in British libraries'. But he thought that the organization of the book around disciplines detracted from an overall picture – possibly of the kind he was attempting in his dissertation on the progress of the modern mathematical and physical sciences for the *Encyclopaedia Britannica*. Instead he suggested a way of translating it into a four-stage general history of science from 1450 to 1850, thus removing Whewell's focus on separate disciplines (J. D. Forbes 1858, 287; see Sarton 1936 for a more direct criticism). Forbes suggested that two

[7] For advice to seek the aid of specialists, see Brewster 1842a, 302. Whewell was criticized for his account of the roles of Charles Bell and François Magendie in the discovery of the sensory–motor division of the nervous system. For his reply, see Whewell 1837b.

generalizations derived from this panorama of science over 400 years. First, that 'great theories' were as rare now as they were in 'the first ages of modern progress'; and second, that 'increasing toil' is necessary as science advances. This latter point had a direct bearing on Whewell's *History* which, in spite of its qualifications, asserted the crucial role of individual genius. However, in Forbes's account, with its interest in long-term developments rather than changes within particular disciplines, this looked like a historical description rather than a methodological lesson. That is, while it may have held for earlier periods, present science seemed to depend on a different dynamic:

more hands must be employed at each successive stage of discovery – and that as the pride of man is flattered by the enlargement of the realm which he calls his own, less and less is achieved by individual prowess, and the more is he indebted for success to the preparatory labours of those who went before, and to the assistance of his contemporaries. (J. D. Forbes 1858, 294)

Whewell's writing on the history of science can also be approached from the perspective of contemporary political historiography. We have seen that he qualified the triumphalism of most contemporary accounts, and this can support the view that the *History* was influenced by Liberal Anglicanism and its critique of Enlightenment assumptions. But this contextualization can be extended by taking up Burrow's analysis of Whig political historiography. This suggests that the disciplinary histories of science which, in their celebrations of progress and their dismissal of past ideas, have come to be seen as the epitome of 'Whiggish' history, were in fact far less subtle than some of the classical Whig political discourse. This betrayed tensions between tradition and progress, continuity and discontinuity, reform and revolution in a manner that parallels Whewell's *History*. Furthermore, by deploying these motifs from general history Whewell effectively related the scientific enterprise to other political and cultural activities. In this sense, his work may be seen as legitimating not particular scientific disciplines, but science itself.

We have now seen how Whewell used the review journals, his Bridgewater treatise, various addresses, and the history of science to discuss the nature of science and the personal qualities of its practitioners. In his major essay review of 1841, Herschel was willing to grant the *History* the status that Mill awarded to Carlyle's account of the French Revolution; but he thought the *Philosophy* went too far in the direction of idealism. However, as the next chapter shows, Whewell believed that the history of science would not in itself vindicate the cultural value of science. This required a philosophical account of its epistemological and moral foundations. Moreover, the chance to supply this was Whewell's means of justifying his role as a metascientific critic.

CHAPTER 7

Moral science

In solitude he [Kant] contemplated his mind with close attention; the examination of his thoughts lent him new strength to support his virtue.

Madame de Staël, *Germany*, 1813

I have been thinking several times about my philosophy; travelling, often alone, one has many happy moments of such speculation.

Whewell to Jones 21 August 1834 WP, c. 51[175]; from Fort William, Scotland

In his Preface to the *Philosophy*, Whewell recalled that Sedgwick had asked for the 'moral' of the *History*, and that he had replied that 'the moral would be as long as the story itself' (Whewell 1840a, I, iii). It is not certain that what Whewell provided in 1840 was the moral Sedgwick had in mind, but there was definitely more than collegiality behind the dedication of this second major work to the Woodwardian Professor of Geology. Whewell presented his new book as a continuation of the fight against the 'fallacies of the ultra-Lockian school' – an enemy that Sedgwick engaged in 1833 in the field of ethics or moral philosophy, in his famous *A discourse on the studies of the university of Cambridge*. Whewell announced the continuation of this battle into the philosophy of the natural sciences (Whewell 1847a, I, iv).

This chapter shows how Whewell's *Philosophy*, the culmination of his thought on the nature and growth of science, can be considered as part of a moral discourse on science. In Whewell's time the term 'moral science' referred to a set of issues associated with the study of man and society, including systematic analysis of human ethical behaviour, and the range of questions raised by psychology, language, anthropology, social statistics, and history, such as the distinction between humans and animals, intelligence and instinct, freewill and determinism – all of which had implications for the concept of human nature. In the *Philosophy*, Whewell was arguing that in another sense, 'moral' science, or good and proper science, depended on a particular epistemology or philosophy of knowledge – one that upheld a Christian view of man's intellectual and ethical character.

In order to recover the connections here, we need to consider Whewell's various projects dating from the 1820s, and to analyse the relationships between them. It was the interaction between these earlier concerns – inductive sciences, political economy, and ethics – that made the *Philosophy* a grand statement on knowledge, embracing both moral and physical science. These links provided a significant vindication of Whewell's metascientific vocation. This approach requires some departure from two existing interpretations: first, one that reads the *Philosophy* largely in terms of the reaction of its critics; and second, one that abstracts it from the relevant social and moral contexts.

EMPIRICISM AND IDEALISM

It has been widely accepted that Whewell's *Philosophy* represented an idealist response to an established British, or at least English, empiricism. A recent essay claims that Whewell was punished for his 'dissent from official doctrine' – an orthodoxy characterized by Baconianism, empiricism, and inductivism. Herschel and Mill, as the guardians of this doctrine, sought to bury the heresy Whewell uttered, partly by ignoring its radical challenge (Wettersten and Agassi 1991, 345–6, 358). But although sympathetic to Whewell's work, the authors are in danger of imposing anachronistic divisions on the early-nineteenth-century debates. The trilogy of terms comprising this orthodoxy did not have stable meanings or connotations in this period. Even in the controversy between Mill and Whewell, which did crystallize some of the dichotomies accepted in later discussions, the clarity of the dispute is less than supposed. For example, there was considerable overlap of opinion on the role of inference in observation – an issue usually taken as a discriminator for wider philosophical differences (Buchdahl 1991). And as argued below, Baconianism was not firmly linked with a definite empiricist epistemology as the natural opponent of Cartesianism.

To avoid such problems, we need to be wary of discussing

Whewell's work solely in terms of the reaction of his critics. There is, of course, a sound precedent for this – namely the confession of J. S. Mill that it provided him with the target he needed:

> During the re-writing of the Logic, Dr. Whewell's Philosophy of the Inductive Sciences made its appearance; a circumstance fortunate for me, as it gave me what I greatly desired, a full treatment of the subject by an antagonist, and enabled me to present my ideas with greater clearness and emphasis . . . in defending them against definite objections, and confronting them distinctly with an opposite theory. (Mill 1971, 133)

No doubt with this in mind, John Passmore once remarked that if Whewell did not exist Mill would have had to invent him (Passmore 1966, 19). But if we approach this debate from the period prior to the *Philosophy* it is possible to suggest that the converse was also true: namely, that in advancing his philosophy of science Whewell raised the stakes by inventing a well-entrenched empiricist epistemology of *science* as the offshoot of utilitarian *ethics* – one which therefore entailed all the moral horrors of its companion.

Let us first examine the conflict between empiricism and idealism in which Whewell's work has been situated. Although this confrontation is now routinely accepted as one of the major philosophical debates of the nineteenth century, and the dispute between Whewell and Mill is seen as a representative moment, it is important not to assume it as a ready-made context of Whewell's work. On the contrary, it is possible that Whewell partly constructed the terms of the debate with which we are familiar. Of course, this is not to say that such an ancient philosophical division was absent from English debates before 1840, but rather to suggest that its definite connection with the epistemology of the *physical* sciences was not clear until after his major works.

It is instructive to compare a number of key reviews of English works on philosophy, in the most general sense. As we know, Whewell and the young Mill reviewed Herschel's *Discourse* in 1831. In 1835 Mill also reviewed Sedgwick's *Discourse*, quite unfavourably. In 1841 Herschel reviewed both Whewell's major books in a sixty-one page article for the *Quarterly Review*. The interesting comparison is that between the emphasis on methodology in the early reviews and its replacement by epistemology as the main controversial issue in the response of Herschel to Whewell. This can be illustrated by contrasting the set of key issues that Mill and Herschel regarded as relevant to the works of Sedgwick and Whewell respectively. Mill felt that Sedgwick had given an inadequate account of the study of nature, suggesting that he could have canvassed 'the methods of the various physical sciences', the differences and similarities in their logic, kinds of evidence, their applicability to other subjects (Mill 1835, 38). None of these were epistemological issues. In contrast, these are some of the questions raised, in Herschel's view, by 'the philosophy of science' as treated by Whewell: 'What is the nature of general and of universal propositions? Are all true universal propositions *necessary* truths, or is any truth, or all truth, necessary?' The series continued for two pages, indicating what Herschel called 'diametrically opposite views of the Philosophy of Knowledge' (Herschel 1841, 180–1).

Herschel stressed that Whewell had taken the discussion of science into an area of 'thorny . . . and abstruse considerations'. As the recent work of Joan Richards suggests, such epistemological problems were more usually raised in early-nineteenth-century debates over mathematics. Whewell was a participant in these because, in his textbooks, he defended a view of mathematical concepts as descriptive of fundamental notions of space and number which had reference to the world. On this view, definitions in geometry, far from being arbitrary, as suggested by Dugald Stewart and William Hamilton, were intelligible because they were grounded in external reality (J. Richards 1988, 20–3; Herschel 1841, 206–7). Richards shows that Herschel, in spite of his earlier campaign on behalf of analysis, agreed with Whewell in stressing the descriptive character of mathematics and then using it as the example of solid truth achieved by human inquiry.

But Herschel's review of 1841 registered a new arena of dispute. Whewell had shifted the debate to *physical* science, affirming an idealist epistemology as an account of how natural knowledge was achieved.[1] Herschel was compelled to respond at this level, explaining the position of an alternative view of science on the 'thorny' question regarding 'the grounds of human belief . . . the very nature of truth itself' (Herschel 1841, 180). This was quite different from the more practical, methodological issues on which he had concentrated in his *Discourse*, and also from those which Mill had mentioned in his

---

[1] As he was composing the review Herschel said that the *Philosophy* had 'led me to bring into question all my own preconceived or habitual opinions and knowledge acquired by reading' (Herschel to Whewell, April 1841, WP, O 15. 46[19]).

review of Sedgwick. Four years later, as President of the British
Association, Herschel seemed to be trying to recover something of the
earlier situation. Acknowledging the epistemological dispute between Whewell's 'peculiar and *a priori* point of view' and Mill's
empiricist position, Herschel advertised what he said was their
agreement on 'the most essential features' of 'the inductive philosophy' (Herschel 1845, xl; Yeo 1989). This was an attempt to
emphasize a consensus over methodology, even though epistemology
was a site of contention.

The reaction against utilitarianism

Before 1840 in Britain the key philosophical contest was mainly
fought over ethics, not science. This was the conflict between
utilitarianism and some version of intuitionism; it extended into
language and was thought to bear on conceptions of human nature
(Aarsleff 1971; Yeo 1979). These were the issues on which epistemological battle lines were drawn; and it was here that the attack
on John Locke's theory of knowledge was focused. When J. S. Mill
later spoke of the 'reaction of the nineteenth century against the
eighteenth', he was referring to this debate, in which Coleridge and
Bentham were the archetypal figures (Mill 1859, 1, iv–v).

One phase of the reaction Mill described was the controversy about
the influence of Locke in the Cambridge curriculum. This came
under scrutiny in the 1830s and was most publicly highlighted by a
sermon delivered by Sedgwick in December 1832 and published as *A
discourse on the studies of the university of Cambridge* in 1833. Whewell
wrote to Sedgwick on behalf of the undergraduates who requested its
publication, and it went through a further four editions, the last, of
1850, being accompanied by a notorious preface and appendix, five
times the length of the original, in which Sedgwick spoke his mind on
transmutationism, materialism, pantheism, and German idealism.
The original *Discourse* had a more limited and local reference: namely,
the content and character of Cambridge undergraduate education in
natural philosophy, classics, and moral philosophy (Whewell to
Sedgwick, 23 December 1832, in Clark and Hughes 1890, 1, 400–5).
This last topic occupied the largest section and was the main reason
for the book's fame, because Sedgwick launched an emotional attack
on the two key texts of the Cambridge curriculum – Locke's *Essay on
human understanding*, and William Paley's *Principles of moral and political*

philosophy. When Mill reviewed the *Discourse* for the *London Review* in 1835 he presented its concerns as somewhat parochial and dismissed its critique of Locke as uninformed. Mill did not seem to recognize Sedgwick's work as part of a widespread disenchantment with Locke's perceived legacy to the nineteenth century; but the acclaim for the *Discourse* was not limited to Cambridge. Thus Robert Southey wrote to Sedgwick thanking him for using 'your sledge-hammer against the Utilitarians; and counteracting the mischief which has been done by Locke and Paley' (Southey to Sedgwick, 10 February 1834 in Clark and Hughes 1890, 1, 427).

Paley's *Principles* of 1785 had been a set text since 1787 (Garland 1980, 57; J. Gascoigne 1989, 243). In this work, Paley questioned the earlier ethical theories of Butler and Hutcheson, arguing that the instinctive moral feelings they discussed could not be sufficiently distinguished from habits and prejudices. He drew on the scenario, common to much Enlightenment thought, of the wild 'savage' 'cut off in his infancy from all intercourse with his species' and predicted that such an individual would have no instinctive perception of right and wrong (Paley 1809, 1, 10–11). The observable variety of morally sanctioned behaviour in different cultures confirmed the absence of a uniform moral sense in all men. Paley concluded that 'there exist no such instincts as compose what is called the moral sense', and proceeded to a system that combined utilitarian ethics and Christianity. Virtue, he said, is 'the doing good to mankind, in obedience to the will of God, and for the sake of everlasting happiness'. Actions were then estimated, in moral terms, by their tendency to produce this result (1809, 12–16, 18, 42, 72).

Sedgwick protested that such a 'utilitarian theory of morals' had to be denounced, not only because it was incorrect, but because 'it produces a degrading effect on the temper and conduct of those who adopt it'. One issue here was the relativist implications of such a theory: if expediency was the measure of right, and everyone claimed liberty of judgement, virtue and vice could no longer have 'any fixed relations to the moral condition of man, but [would] change with the fluctuations of opinion' (Sedgwick 1834, 63). Sedgwick was appalled by the references in Locke and Paley to cross-cultural diversity of ethical ideas. Since university education was aimed at the morals as much as the minds of undergraduates, Sedgwick believed that an urgent reform of moral philosophy in the curriculum was needed. But he contended that this would not be possible without a rejection of the false Lockian view of human nature underlying Paley's approach to ethics.

The significant point for this discussion is that in 1833 Sedgwick saw no potential moral problems in the use of the natural sciences at Cambridge. He praised Paley's other famous work on *Natural theology* as an example of the successful combination of good science and teleology, and made no association between physical science and Locke's epistemology. He did not contend, as Whewell did later, that Locke's theory of knowledge was the basis of a rival, non-theistic view of *physical science*. Similarly, in reviewing the *Discourse*, although Mill defended a utilitarian ethical theory (not necessarily Bentham's) against 'intuitional' moral philosophy, he did not advance an empiricist conception of science against that presented by Sedgwick.

Method versus epistemology

This situation differs markedly from the well-developed oppositions between Mill and Whewell which attracted the attention of philosophers of science in the 1960s. One of the reasons for this difference relates to the Baconian framework in which science was discussed in the early nineteenth century. In Britain, the Baconian legacy was read as mainly methodological rather than epistemological in focus. In France, on the other hand, Bacon had been adopted by Diderot and D'Alembert as the patron saint of the Enlightenment, and 'Baconisme' became a polemical platform on knowledge and its function in society. In 1740 in his *Essay on the origin of human knowledge*, Condillac identified Bacon as the first to stress the genesis of all knowledge in sensory experience, and by the early nineteenth century his work was seen as the ancestor of the sensationalist philosophies of Condorcet, Cabanis, and de Tracy. When Kuno Fischer sought to introduce Bacon to a German audience in 1856, he presented him as the founder of a philosophical tradition manifested in the writings of Hobbes, Locke, and Hume (Yeo 1985, 256).

I have argued elsewhere that in early-nineteenth-century Britain, Baconianism was not associated with these broader philosophical and political interpretations, and the French image of Bacon as the patron of a wide-ranging social philosophy was effectively neutralized (Yeo 1985). Instead there was a heavy emphasis on Bacon as a methodologist, particularly in the writings of the Scottish philosophers, in a manner that avoided or sidestepped direct discussion of epistemology.

Furthermore, there was no common linkage between Bacon and Locke in this period. Bacon was not automatically identified with an empiricist theory of knowledge. In fact, some of his readers interpreted him as favouring a position that conflicted with empiricism. On the one hand, Thomas Brown told Macvey Napier that Bacon's philosophy was marred by its admission of forms or 'essences' in nature; on the other, Coleridge referred to Bacon in 1817 as the 'British Plato' (Coleridge 1817, 27; Yeo 1985, 257). Against this background, it is not surprising that as late as 1865 David Masson was unsure about the link between Bacon and empiricism (Aarsleff 1971, 403).

The most influential article on Bacon by Macaulay in the *Edinburgh Review* of 1837, did not cast him as a philosophical empiricist, although it criticized both the pretensions of his methodology and the virtue of his character (Yeo 1985, 271–2). Indeed, Macaulay did not use the term 'empiricism' or any of its variants. It is important to realize that in this period the word 'empiricism' was more often used to suggest a preference for fact over theory, rather than any precise epistemological position. Thus a writer in *Blackwood's Magazine* in 1843 criticized the anti-theoretical bias of English thought and castigated 'empirics' who relied solely on practical first-hand experience (Grove 1843, 517). This usage was also employed by Mill: in 1836 he declared that if an art were not based on scientific principles it would not be 'philosophy, but empiricism' (Mill 1836, 4).

There were certainly negative assessments of science which implicated Bacon. His plea for utility was increasingly seen as coming to fruition in the industrial and technological successes of the century, many of which were attributed to science. There were then criticisms of this tendency, particularly from Romantic writers such as Wordsworth, Shelley, and Carlyle, but also, as we shall see, from conservative High Church theologians. Although these critiques stemmed from different agendas, there were some recurring themes. The novelty and visibility of science and its technical applications were distracting attention and prestige from the moral and metaphysical sciences. Carlyle gave this complaint epic quality in his essay on 'the signs of the times' in the *Edinburgh Review* of 1829. Here he characterized contemporary attitudes as typical of a 'Mechanical Age' in which a fixation on material and practical knowledge militated against mental and moral philosophy. Approaches to these topics which still existed were themselves dominated by mechanical and atomistic concepts derived from the physical sciences (Carlyle 1829, 442–6). In 1838 another commentator warned that Britain was 'sinking ... into natural and mechanical philosophy' while Europe was cultivating exciting studies in the philosophy of history and art (Pusey 1838, 462). The Catholic theologian, Nicholas Wiseman, agreed with these observations, predicting that there was a

real danger of seeing the next generation brought up in the ideas of many of the present, that man is a machine, the soul is electricity, the affections magnetism, that life is a rail road, the world a share-market, and death a terminus. (Wiseman 1853, III, 591)

Some writers linked Bacon with this, claiming, as the *Quarterly Review* did in 1838, that he saw knowledge as utility and power rather than as something to be pursued for its intrinsic value (Pusey 1838, 502, 505). But this was still not explicitly connected with any specific critique of an inadequate philosophy of science; Bacon was rarely attacked for espousing an empiricist epistemology. Whewell was the first to do this in a major way, and this helped Mill to make the subsequent public debate turn around idealist versus empiricist accounts of science.

Mill knew that an account of science was crucial to arguments on other topics. In his autobiography he recalled that the chief strength of the intuitionist school – in which he placed Whewell and Sedgwick – was that their views on morals and politics were supported by appeals to the 'evidence of mathematics and of the cognate branches of physical science'. A victory against this position would have to be won in these subjects, for 'to expel it from these, is to drive it from its stronghold' (Mill 1971, 134–5; see Long 1841, 8, for the view that Whewell's scientific reputation assisted his campaign against utilitarian ethics). But Mill admitted that even after the work of Bentham and James Mill, his father, there was no effective counter to the 'intuition philosophers' and their account of necessary truths in mathematics and science. Indeed, his own essay on 'Bentham' in 1838, seemed to underline this absence. He recounted the utilitarian view of law, politics, psychology, and morals but did not mention physical science. However in 1840, when he discussed Coleridge as the representative thinker of the opposing 'transcendental' philosophy, he noted that there was a conflict respecting 'the sources of human knowledge'. Now he included a reference to science, explaining Coleridge's rejection of 'Locke and his followers': 'Even science, it is affirmed, loses the character of science in this view of it, and becomes empiricism; a mere enumeration and arrangement of facts' (Mill 1980, 109–11).

On Mill's survey of the early-nineteenth-century intellectual scene, there was no extensive articulation of the idealist philosophy of science he wished to oppose. Before 1840, when Whewell spelt out such a position, the 'intuitionist school' had the field to itself; its members had not been forced to elaborate and defend their assumptions. After reading Whewell's *History* in 1837, Mill set out to offer an account of induction within an empiricist framework in his *System of Logic* (Mill 1971, 124).[2] While he was writing this book, Whewell's *Philosophy* appeared and, as Mill recounted, his task was thus made easier since the *Philosophy* put science at the centre of epistemological argument, claiming that a Lockian or sensationalist theory of physical science was erroneous and morally dangerous.

Intellectual context of Whewell's project

It may now be possible to form a picture of the context in which Whewell's concerns developed. We have seen that he did not have to combat an established empiricist philosophy of science such as that associated with Bacon in France. Before the appearance of Comtism as a significant movement in Britain[3] – that is, before Mill's *Logic* – the precise embodiment of what Whewell called 'Lockian' errors, in the philosophy of *physical science*, is not so easily located as is often supposed. Although both Brewster and Powell reviewed Comte's *Cours de philosophie positive* (in 1838 and 1839) and warned that it failed to draw the appropriate theological inferences from the results of science, neither regarded his work as having epistemological implications for current views of scientific knowledge. Earlier, Herschel affirmed a version of Baconianism in his *Discourse*, but he did nothing to promote epistemological controversy. Where he touched on such issues Herschel allowed a space for *both* rationalist and experiential components of scientific knowledge, almost referring to two branches of natural philosophy – abstract/mathematical and empirical/experi-

---

[2] Mill acknowledged the 'aid derived from the facts and ideas' in the *History*. See Mill 1856, i, vi.
[3] For the rise of the English Comtism from this period, see Kent 1978, 56.

mental (Yeo 1989, 543; Fisch 1991a, 30–1, 36). As noted in chapter 4, neither of the two major reviewers of Herschel's book – the young Mill and Whewell himself – interpreted it as relevant to a conflict between empiricism and idealism. As late as 1836 Whewell asked Jones to put on paper the case for 'experience' as the basis for the 'simplest mechanical truths', explaining that he was 'desirous of getting this opinion in its best and most definite shape, because the negation of it is a very leading point of my philosophy' (Whewell to Jones, 6 September 1837, in Todhunter 1876, II, 259). That is, Whewell needed Mill's *Logic* before it appeared.

The perspective of Baden Powell, who studied the philosophical writings on science of the 1830s, is revealing. During this period, Powell read Whewell's position as one in the debate conducted by Stewart and the Oxford Noetics on the nature of mathematical and scientific knowledge. Stewart distinguished between the axiomatic knowledge attainable in geometry and the experiential knowledge of the physical sciences. In geometry, the model of perfect science, deductions could follow from fundamental axioms and could produce necessary and universal truths. But in the physical sciences which relied on inductions from experience – observation or experiment – there could be no such certainty in the conclusions reached (Corsi 1988, 151). Powell saw that Whewell was departing from this position, also upheld by his Oxford colleague, Richard Whateley, by claiming that there were a priori axioms in mechanics, even though they were unfolded by experience (Powell 1837, 10–11, 59–60; Powell to Whewell, 29 January 1838, WP, Add. MS. a. 210[170]). The relevant point here is that, in Powell's view, the position Whewell was attacking was not that of a morally dangerous empiricist epistemology of science, but rather that of Oxford intellectuals who, like Whewell and himself, were seeking to harmonize science with Christianity. In short, then, the most specific discussions of science available in the 1830s did not espouse the kind of endemic errors Whewell referred to in the preface of the *Philosophy*, where he claimed to be taking Sedgwick's campaign against Locke into another territory occupied by the enemy.

Later in this chapter we will see that during the 1830s Whewell came to realize that an adequate defence of science required an epistemology that actively disrupted any association between science and Utilitarianism. But given the previous discussion, it is likely that he was replying to the negative appraisals by the Romantics and High Church critics, who branded science with such a link, rather than to any positive programme conducted by the utilitarians.

THE MORAL OF WHEWELL'S PHILOSOPHY OF SCIENCE

Another influential perspective on the *Philosophy* is that initiated by Todhunter: namely, that it marks the watershed between Whewell's serious attention to science and the transition of his work to moral philosophy. This has often joined with interpretations eager to jettison the moral and theological attachments of his work in order to insert it in twentieth-century philosophical agendas. In the remainder of this chapter I argue that moral concerns were present in Whewell's earliest reflections on the nature of science, and were later crucial to the legitimation of his idealist epistemology.

We have seen that in dedicating the *Philosophy* to Sedgwick Whewell said it contained the 'Moral' of the story previously told in the *History*. In reviewing both of these works in 1841, Herschel said that they raised grave questions because 'there are duties and responsibilities, individual and social, attached to their discussion' (Herschel 1841, 180). What all three contemporaries agreed on here was that a correct understanding of the nature and practice of science was a moral issue. Whewell was arguing that his account of science, involving a particular epistemology, was morally preferable to the alternative he attacked: namely one founded on a Lockian theory of knowledge.

The sense in which a work on the method and epistemology of science can be said to involve moral issues is one that has to a large extent been lost. Yet Herschel was by no means unusual in believing that human moral and intellectual capacities were closely related, in that the performance of responsibilities depended on rational faculties and, conversely, the right exercise of reason was a moral duty. On his reading, Whewell's *History* made science itself an object of inductive inquiry and revealed social, psychological, and intellectual factors that either facilitated or hindered scientific advance. While he agreed with Whewell that it was impossible to lay down infallible rules of discovery, Herschel insisted that such inquiries into the history and philosophy of science were profoundly important because they touched on man's responsibility for the proper use of Divinely given intellectual capacities. It followed from this that epistemological theories about the source of ideas and concepts, and methodological formulations about the rules of inference from evidence or procedures of inquiry, had moral connotations.

Whewell and Herschel were not alone in believing that the manner in which human beings employed their intellectual powers was a moral issue. This topic was specifically addressed in works such as John Abercrombie's *Inquiries concerning the intellectual powers and the investigation of Truth* of 1830 – which went through ten editions in ten years – and his aptly titled *The culture and discipline of the mind.* These analysed modes of observation and inference, making it clear that the proper use of the Divine gift of reason was not only a philosophical issue but a moral duty. Similar convictions informed the discourse of natural theology, in which writers claimed that a pure mind and the observance of proper methods of inquiry guaranteed the perception of design in nature and the inference of an intelligent Deity. Thus in his *Remarks on scepticism* Thomas Rennell ascribed all heterodox opinion to the deforming effects of pride and immorality upon the natural perceptions of reason; Abercrombie referred to 'a previous moral corruption of the mind' (Rennell 1819, 15–20; Abercrombie 1837, 41; Yeo 1977, ch. 3).

This conjunction of theology and epistemology also appeared when natural theologians extended the design argument to the moral sphere, thus showing that the intelligent Creator of the material world was also the providential 'Governor and Judge of men', as Whewell put it in his Bridgewater treatise of 1833 (Whewell 1834a, 254). In demonstrating the adaptation of the mind to the world, many writers asserted that the ability of human intellect to know the world was in itself proof that God had adapted one to the other. The Scottish theologian Thomas Chalmers adduced the consonance between certain abstract mathematical ideas and the structure of the world, asserting that this indicated the 'intervention of a Being ... who had adjusted the laws of matter and the properties of mind to each other' (Chalmers 1836–42, II, 159). Similarly, for Powell, man's success in comprehending nature displayed the adaptation between the intellectual and physical spheres of Divine creation; it completed 'an essential part in the universal harmony' of the world (Powell, 1838, 203). Herschel summarized this powerful notion in his poem, 'Man, the interpreter of nature', affirming that the universe was incomplete before the appearance of man and his ability to comprehend the harmony of nature:

Man sprang forth at the final behest.
His intelligent worship
Filled up the void that was left.
Nature at length had a soul

Herschel 1857, 737; Yeo 1979, 498

Awareness of this context is crucial for understanding Whewell's major works. Much early scholarship regarded such a mixture of theological and epistemological concerns as a peculiar characteristic of Whewell's idealist philosophy, which was seen as outside the mainstream of contemporary British thought, and certainly foreign to prevailing approaches to science (Blanché 1967; Seward 1938; Belsey, 1974). However, it should now be clear that this combination was present in natural theology and in the writings of other scientific commentators such as Herschel, Brougham, Brewster, and Powell. In fact, in taking exception to Whewell's epistemology, Herschel offered an alternative account which also argued within a theological framework, suggesting that scientific knowledge was possible because God had endowed man with the capacity of drawing analogies from experience, rather than from innate fundamental ideas (Herschel 1841, 182; Wilson 1974). If Herschel and Whewell can be seen as advancing empiricism and idealism respectively, then this was a debate in which theology, and morals, appeared on both sides.

Whewell's *Philosophy* brought two senses of 'moral science' together. It concerned the best philosophical account of both human nature – its intellectual and moral faculties – and the science produced by men in their use of the Divinely bestowed gift of reason. This is why he saw nothing strange in publishing the *Philosophy* while occupying the Knightbridge Chair of Moral Philosophy. But it is important to understand that what Whewell did, although it can be seen in the context of an established discourse on the proper use of reason, was not an obvious strategy: it did not meet with the full support of his friends and colleagues, even though they praised the broad scale of his endeavours. Jones told Herschel that he hoped Whewell might be dissuaded from his idealist position (Jones to Herschel, 1 June 1841, HP, vol. 10, no. 370). De Morgan commented that it was strange to find 'that the doctrines of Kant and Transcendental Philosophy are now promulgated in the university which educated Locke' (De Morgan 1840, 707). And another reviewer worried about the attempt to 'popularize the *whole* of Kant in the cloisters of Cambridge' ([Butler] 1841, 201). We need to understand why Whewell contended that his anti-Lockian epistemology was morally sound, crucial to the reform of philosophy, and essential to a proper account of science. (For more on Powell's continuing disagreement, see Powell 1849 and 1850; also ch. 9 below.)

This requires awareness of the range of discussions in which Whewell was involved from the 1820s. These can be listed as: (1) the attempt to define the nature of inductive science, which he confided mainly to Jones, and which Fisch has connected with his textbooks on mechanics; (2) correspondence with Jones on political economy and inductive method in the moral sciences; (3) a long-term but less intensive correspondence with Rose and Hare on general issues of philosophy, German thought, poetic criticism, and the need to formulate an alternative system to the ethical theories of Locke and Paley. As noted earlier, Todhunter claimed that 1840 marked a watershed in Whewell's intellectual focus, a shift from science to moral philosophy. But a survey of these three areas indicates Whewell had been concerned with *both* these subjects for at least fifteen years.

Whewell's early philosophical reflections

Fisch has recently argued that Whewell's early textbooks were the starting-point of an attempt to explain the structure and development of excellent physical science, but that in the course of this quest his assumptions and key questions underwent a crucial shift. The central topic of the *Philosophy* – in Whewell's words, the 'Fundamental Antithesis of Philosophy', or the dialectical relation between sensations and ideas in all knowledge – was not present at the start of his inquiry into the inductive sciences. Whewell's mature theory was 'not born of epistemological controversy'. In Fisch's reconstruction the original focus was on method, and the concern with epistemology only took definite shape after about 1834 (Fisch 1991b, 65 and 1991a, 106–8). If we accept this analysis we need to understand how the contrast between empiricism and idealism came to be the *Leitmotiv* of the *Philosophy*, with its strong moral reforming tone, aimed at Lockian and sensationalist accounts of science. The clue to this transition may lie in Whewell's other concerns – political economy and ethics – and the interaction between them.

Although the early inductive notebooks do not contain epistemological speculation, such issues were present from the start of

Whewell's correspondence with Rose and Hare. His relationship with these two non-scientific intellectuals can be seen as the complement, and perhaps counterweight, to his important relationship with Jones and Herschel. Their discussions, as noted in chapter 3, moved across a wide range of subjects, from poetry to science, but there was a continuing concern with the grounds and sources of knowledge. Whewell was at first sceptical about the value of systems of poetic criticism and metaphysics, but this is not to say that he was indifferent to such issues – his reading notes illustrate a critical engagement with Locke, Stewart, Brown, and Kant (see Todhunter 1876, I, 345–7 for Whewell on 'personal identity'). However, he did not seem to be committed to any of the major schools of philosophy. Thus when Rose wanted to know 'how you modern philosophers' react to the claim that 'the doughty Scotchmen's remedy against Berkeley and Hume is not worth a farthing', Whewell did not push any particular position but raised queries about the terms of the debate:

The important question is, whether the objects we perceive are independent of us in their relations and sequences, and of that we have complete evidence as far as the proposition is intelligible ... But I will tell you what the mischief is ... People choose to ask whether the objects we perceive are *without* the mind or not. What do they mean? I know what is meant by a church steeple being on the outside of the eye, or a dead dog on the outside of the nose ...; but in point of fact the relation between the mind and exterior objects is not one that can be expressed by any such beggarly speech, it is that of *perceiving* and *perceived*. (Whewell to Rose, 22 September 1822 in Stair-Douglas 1881, 78–9; Rose to Whewell, 2 September 1822, WP, Add. MS. a. 211[133])

A year later he voiced another complaint about the way in which Rose and Hare, following Wordsworth and Coleridge, had constructed a philosophical opposition between the rational and the affective. This was part of his resistance to the set of dichotomies deployed by the Romantics in which science was associated with the rational and pragmatic and contrasted with imagination (see ch. 3). But if Whewell was not yet inspired by Coleridge, he was already dissatisfied with Locke. As early as 1814 he told his former schoolmaster, George Morland, that he was reading the *Essay* in bed, but then 'grew out of humour with Locke' (April 1814, in Todhunter 1876, II, 2). His notebook and diary have entries suggestive of the critical stance taken in his later letters to Rose. For example, on 6 August 1817: 'Locke's Essay book 1 – No innate principles. The whole question turns upon what we understand by innate principles' (WP, R 18. 9^2, p. 10).

But in these early criticisms there is no indication that Whewell saw *moral* consequences in this Lockian philosophy. However, at least by 1821, he was reading Madame de Staël's *De l'Allemagne* of 1810, noting its account of Kant's transcendental categories, but also, his ethical views. Significantly, the section of this book dealing with Kant is headed 'Philosophy and Morals', underlining the connection de Staël wanted to suggest (Staël-Holstein 1813, III, ch. 1). Whewell recorded that:

German theory of Morals – opposed to system of utility – Kant makes the principle of duty (a necessary feeling?) the foundation of morals ... the rules of duty absolute – the prospect of a future life should have no influence. (WP, R 18. 9^8, 14 February 1821, p. 17)

By 1825 Whewell had begun to take notes from Kant's *Kritik der reinen Vernunft*: 'We possess a priori synthetical judgements ... The *Problem* then is *How* a priori synthetical judgements are possible' (WP, R 18. 9^{13}, 18 December 1825, p. 19). However, this interest in Kant had certainly not translated into a better view of Coleridge: Whewell noted on 25 July 1827 that he doubted if Coleridge 'fully understands Kant' (WP, R 18. 9^{14}, 25 July 1827).

On Fisch's account, this reading of Kant had not yet influenced Whewell's view of induction; but it may well have affected the way morals and philosophy interacted. Rose and Hare were probably crucial in convincing Whewell that Locke's epistemology was a religious and moral danger, thus encouraging him to take such questions more seriously than he had done before 1821. While they were writing to Whewell, both were aware of Coleridge's deep antagonism to Locke. Coleridge had written to Rose, attacking Locke and Condillac (Coleridge to Rose, 17 September 1816, in Griggs 1956–71, IV, 670). His *Biographia literaria* of 1817 was critical of sensationalist philosophy and, as early as 1805 he had linked this with 'the vile cowardly selfish calculating ethics of Paley, Priestley, [and] Locke' (Aarsleff 1971, 402). Rose endorsed this critique and, in reply to Whewell's expression of frustration with metaphysics, underlined the moral and religious implications of Scottish and Lockian philosophy. It was absurd of defenders of the 'Scotch school' to talk of its 'connexion with Religious Matters as if both it and the Locke System did not alike lead to the most hopeless Scepticism as to the existence of a first Cause'. Both produced a view of human mind as if it were 'the same sort of thing as Babbage's Calculating Machine' (Rose to Whewell, 5 October 1822 WP, Add. MS. a. 211[134]). Whewell heard a similar denunciation from Hare in the same year, conducted through ridicule of a favourable comment on Locke in the *Quarterly Review*. Hare related that the reviewer:

looks back to the reading of Locke's Essay as an Era in his life, that it was the creative let there be light which dawned upon the Chaos of his mind, forgetting that his mind is only one of those puzzle maps which any ... child can put together. (Hare to Whewell, 30 April 1822, WP, Add. MS. a. 77[126])

This naive acceptance of Locke had encouraged the view that 'a Junior Soph now knows much more of the nature of the human mind than Plato and Pythagoras, ... Leibniz and Mallebranche'. He spelt this message out in the next letter, claiming that 'Plato is worth ten thousand Aristotles and 100,000 ... Lockes' (Hare to Whewell, 13 May 1822, WP, Add. MS. 77[127]).

Thus by the time Sedgwick preached his sermon of December 1832 against Paley and Locke, Whewell was already aware that a thorough criticism of utilitarian ethics would have to involve epistemology. But Sedgwick's *Discourse* could not have been a sufficient stimulus for Whewell's extension of this anti-Lockian theme into the account of inductive science. Sedgwick did not suggest that there was a Lockian view of science in need of reform. Like Whewell's defence of science in the 1827 sermons, it was still conducted through conventional natural theology. Whewell told Jones that the main aim of his own sermons was 'to make science fall in with a contemplative devotion, which I don't think was difficult though people seem, from the notion they had of scientific men, to have thought it must be impossible' (Whewell to Jones, 26 February 1827, in Todhunter 1876, ii, 82–3). Sedgwick did not add anything to this mode of affirming *science*, although he did repeat it within the scope of a broader topic. The question, therefore, is, how did Whewell come to believe that this defence was insufficient? One possibility is that his discussion with Jones on political economy, together with his awareness of the debate on ethics, may have suggested the need for a more wide-ranging, philosophical account of inductive science.

Political economy and morals

Whewell and Jones began to discuss political economy in 1822, well before Whewell began his intensive inquiry into the nature of induction. But it is likely that the themes of the early discussion had a subtle influence on the final form of the inductive project. In August 1822 Whewell, almost inadvertently, highlighted the special nature of political economy when on the one hand he said that it involved 'no peculiar principles of observation or deduction', but also noted that it often assumed something about the 'moral and intellectual qualities of man' (Whewell to Jones, 16 August 1822, in Todhunter 1876, II, 49). Boyd Hilton has recently established how closely debate on political economy was informed by religious and moral consider-ations, as the intellectual leaders of the Church sought to formulate a 'Christian Economics' in the course of responding to the different positions of Malthus and Ricardo. Both Hilton and Corsi have shown how these issues divided Oxford intellectuals such as Whately and Nassau Senior from Whewell and Jones at Cambridge. As mentioned in chapter 4, the Cambridge men regarded this dispute as one in which the notion of *good* science was at stake. In their opinion, this had two dimensions: proper inductive method *and* proper conceptions of human nature and God's design in the social sphere (Corsi 1987; Hilton 1988, ch. 5; Yeo 1991a, 181–4).

In reviewing Jones's work on *Rent* in the *British Critic*, Whewell was aware that the Ricardians claimed the title of 'science' for their brand of political economy – one whose doctrines were unpalatable to the landed classes and the established Church. Clerical writers such as Copleston and Chalmers who approached the new science were troubled by its potential for 'secular contamination' and its tendency to present man as 'a blind part of a machine' (Hilton 1988, 50–1). There had been earlier attempts to bring the theories of political economy within the framework of natural theology – for example, John Bird Sumner's *Treatise on the records of creation* of 1816 outlined the Divine moral plan behind the Malthusian laws of population – but when Whately assumed the Drummond Chair at Oxford in 1830 he still felt the need to legitimate the subject (Rashid 1977). It is therefore significant that Whewell contemplated tackling the evi-dence of 'benevolent design in the moral frame of society' in his sermons of 1827 on the religious value of the *physical* sciences (Whewell to Jones, 10 December 1826, in Todhunter 1876, II, 79). In the end he did not deliver this extra sermon; and his subsequent correspondence with Jones suggests that he realized that political economy involved problems which could not easily be resolved within traditional natural theology (Whewell to Jones 26 December

1826 and 26 February 1827, in Todhunter 1876, ɪɪ, 82; also ɪ, 330).

Whewell agreed with Jones that the leading doctrines of political economy – which its proponents claimed as scientific – were morally degrading. His last sermon referred to the existence of general moral laws as necessarily 'present and immediate' in the mind of God, but he warned that some of the current theories, which 'profess scientific validity', such as those concerning the tendencies of population growth, were being interpreted in ways that conflicted with the conviction of God's beneficial design. One answer to this was that some political economists had ignored the lessons taught by the progress of physical knowledge: namely, the danger in pursuing the deductive consequences of just 'one or two assumed principles' without constant reference to relevant observations. In the case of political economy, this meant that secure knowledge could only be gained with 'a recurring reference to individual cases, to conscious feelings, to limited rules' (5th sermon, WP, R 6. 17[17]). As noted in chapter 4, Whewell and Jones developed this methodological critique against the premature deductivism of the Ricardians, arguing that political economy could only be a science if it adopted the inductive process, moving from careful collection and classification of facts to cautious generalizations. This is where Jones was supposed to play his part, providing examples of the variety of economic situations, thus undermining the facile axioms of the Ricardians.

But the failure to abide by proper scientific method was only part of the deleterious influence of the reigning political economy. Whewell and Jones contended that the deductive school had assumed an immoral and unchristian view of human nature. The Cambridge and Oxford writers differed over the relationship of ethics to political economy. Whately was content to stay with Paley's dependent theory of ethics, and while preparing his introductory lectures told Senior that he was 'thinking ... of making a sort of continuation of Paley's *Natural theology*, extending to the body politic some of his views as respecting the natural' (Checkland 1951, 57; Hilton 1988, 53). As mentioned above, this was a strategy Whewell had tried in 1827 and soon abandoned. Unlike Whately, Whewell believed that questions of ethics were central to political economy as a discipline dealing with human motivation and moral behaviour. In 1831 Jones reported that Whately, following Senior, had defined man as 'an animal that makes exchanges', claiming that 'it is in that view *alone* that man is contemplated by political economy' (Jones to Whewell, June 1831,

WP, Add. MS. c. 52³⁸). Whewell replied that he was 'quite ready to fight', insisting that sound views of man's moral nature could only be discovered by inductive, not deductive methods. In his view, Whately was committing the same error as the utilitarians – assuming that 'principles of action are known by consciousness and do not require detailed observation' (Whewell to Jones, 15 July 1831 in Todhunter 1876, II, 123). Jones spelt out the moral results of the error in the preface to his book, remarking that the economic doctrines he was combating were 'first insisted on with a dogmatic air of *scientific* superiority, [then asserted as] an apparent inconsistency between the permanence of human happiness, and the natural action of the laws established by Providence' (Jones 1831, xi–xix). Sedgwick later touched on this theme in his *Discourse*, castigating all political systems 'deduced by *a priori* reasoning from assumed simple principles without comprehending all the great elements of man's moral nature' (Sedgwick 1834, 86). This reiterated the conclusion reached by Whewell and Jones in their critique of deductive political economy.

All this suggests that the issue of morals in connections with science was discussed with specific reference to political economy *before* Whewell claimed that moral and epistemological issues were a general problem for all science. By the time Sedgwick's *Discourse* opened the public campaign against Lockian ethics, Whewell had seen the danger of an association between political economy – a subject claiming scientific status – and utilitarian views of human nature. He may well have realized that such an association would undermine the apologetics of his 1827 sermons – the attempt to harmonize science and religious values. But the concern about utilitarian ethics did not have any logical implications for Whewell's project on the *methodology* of induction, as long as he assumed a distinction between the moral and the physical sciences.

Such a distinction was part of Whewell's position in 1831. In January, when referring to his own papers on the application of mathematics to political economy, he warned the Ricardians that the 'laws of space and number' were of little help in dealing with the 'moral elements of our nature' (Whewell 1831b, 87; also 1831f). Reviewing Jones in July he stressed that the social meaning of rent, taxes, wages, and capital should not be concealed by glib quantitative definitions (Whewell 1831d, 56–7). In the same month he wrote in his notebook on induction under the heading 'conjectures concerning the Domain of Inductive Philosophy'. Here he speculated about the possibility of extending the method of induction to the 'facts of consciousness', noting that although Hume had proposed this, there had not yet been any systematic application of Bacon's method to this realm. At this stage Whewell referred to two kinds of science – 'objective' and 'subjective' – and saw political economy as interesting because it was a borderline case:

Political Economy stands near the boundary line of these two departments of objective and subjective sciences: many of its facts are external (statistical and commercial details), some internal (motives, happiness and virtue, sin[?] and misery). (WP, R 18. 17¹⁵, p. 46, dated 2 July 1831)

Less than a year later, Whewell's discussion of this issue appeared to take a different form. In a letter to Jones of February 1832 about the doctrines of the Saint Simonians, he acknowledged that they had 'several right notions about the character of science', most importantly, that of *'conceptions* which must exist in the mind in order to get by induction a law from a collection of facts; and the impossibility of inducting or even of collecting without this' (Whewell to Jones, 19 February 1832 in Todhunter 1876, II, 141). Apart from indicating a crucial step in the break with orthodox Baconianism, it is significant that in this and other passages from this date, Whewell began to dilute the earlier dichotomy between 'objective' and 'subjective' sciences. He told Jones that when this approach to induction was developed, something which the Saint Simonians had not done, 'we shall have a much clearer view of the nature of general truth and the way of getting at it – including political and perhaps something of moral truth' (Todhunter 1876, 141). By August 1834, when he informed Jones of the likely form of both the *History* and the *Philosophy*, Whewell was even more specific about what he called the 'analogy' between the physical sciences and 'our knowledge respecting morals, taste, politics, language, and generally, all hyperphysical knowledge' (Whewell to Jones, 21 August 1834 in Todhunter 1876, II, 186). It is clear that he now saw substantial common features in what he formerly regarded as different kinds of knowledge. Both 'objective' and 'subjective' sciences depended, as he explained to Jones, on 'certain mental bonds of connexion, ideal relations ... sciential conditions, or what ever else you can help me call them: they are what I called *Ideas* in my last letter'. Furthermore, he concluded that this line of thought had definite implications for morals:

I will add that when I have shewn, as I hope to do, that the relations of duty and of the affections are as fundamental a part of man's thoughts as the relations of time and space, and direct his will in the same inevitable manner, I think I shall have assigned man a moral and social constitution on firm grounds. (Whewell to Jones, in Todhunter 1876, II, 187–8)

Morals as part of the inductive project

In the *Philosophy*, Whewell said that English writers were led to see Lockian errors through 'inquiries into the foundation of Morals' (Whewell 1847a, II, 308). From what has been said so far, this surely applies to Whewell himself. The precise steps of this process are difficult to recover, but it is striking that from 1832, when he began to discuss his new view of induction, Whewell invariably mentioned its relevance to the question of moral philosophy. We know that he admired the undergraduate essay by Thomas Birks (a future incumbent of the Knightbridge Chair) on *The analogy of mathematical and moral certainty* of 1833. Birks argued that knowledge in both these areas depended upon intuitive laws of thought as well as information from the senses. Drawing upon Coleridge, he asserted that there was no reason why 'the evidence of morals should be placed on a lower footing, or regarded as less purely demonstrative than geometry' (Birks 1834, 12, 16).[4] This clearly resonated with Whewell's reflections on the analogies between physical and moral science, and he told Hare that the 'philosophy was most profound and consistent' (Whewell to Hare, 25 December 1833, in Todhunter 1876, II, 175).

As Knightbridge Professor from 1838 Whewell developed this mode of securing the status of ethical knowledge by comparing it with physical knowledge (see Yeo 1991a, 185–7 and ch. 9 below). But this venture into moral philosophy was not a shift away from his work on science, as Todhunter suggested; rather, it is clear that from 1832, Whewell regarded induction and morals as part of the one inquiry. In 1835 he spoke of them as concurrent projects, telling Jones that in James Mackintosh's *Dissertation on the progress of ethical philosophy* of 1830, he had 'got a glimpse, which I have long been wishing and struggling for, of the inductive history of ethics' (Whewell to Jones, 9 May 1835, in Todhunter 1876, II, 212). Only days earlier he had

---

[4] Wordsworth clearly regarded Birks as a student of Whewell, praising Whewell's Bridgewater treatise and Birks's sermon, as if they were connected. See Hill 1978–9, V, 709.

assured Jones that his current reading of Butler, Hobbes, and Mackintosh was not a deviation from 'my proscribed course of writing about Induction':

I shall expect to manage so that what I do will come into its place at some time or other; for the beauty of my Induction is, that it is like the Devonshire man's pie, into which he puts every thing which he catches. (Whewell to Jones, 2 May 1835 in Todhunter 1876, II, 210–11; also 26 May 1835 and 21 August 1835, II, 213–15, 222)

In 1836, when Whewell edited a more accessible edition of Mackintosh's *Dissertation*, he was able to assume that the deficiencies of 'Paley's mode of treating the moral sense' were widely accepted. Now the task was to clarify a new theory that acknowledged the 'independent *existence* and the *supremacy* of the conscience or moral faculty'. Mackintosh showed the way, because in answering the utilitarian appeal to the cultural diversity of ethical beliefs, he stressed the close agreement among men of all races on fundamental moral categories, and on the very distinction of right and wrong (Mackintosh 1836, 13, 20, 62–7). Whewell was impressed by Mackintosh's distinction between the 'theory of Moral sentiments, and the Criterion of Morality', since this allowed the existence of moral sense in man without necessarily rejecting the principle of utility as part of a deductive system of ethics. Mackintosh also offered an answer to those who claimed that all moral feelings could be accounted for in terms of 'the association of ideas', such as the connection of certain acts with either pleasure or pain. As Whewell explained, 'association' was 'employed in the education rather than in the creation of our moral sentiments', and this principle of association presupposed 'laws and powers of the mind itself, according to which the conjunction produces its results'. He suggested that this view corrected the common image of the mind as a blank sheet prior to experience, and supported Sedgwick's image of a piece of paper 'prepared' by a Divine hand (1836, 24–5, 35–6). Mackintosh's answer to association-ism allowed the concept of a moral faculty or principle that was irreducible, rational, and intrinsic to the nature of man (for his support of Butler as another answer to utilitarian ethics, see Whewell 1848a, x–xiv).

The interaction between Whewell's thinking on science and morals was now clearly visible. In introducing Mackintosh's work Whewell employed language and concepts which were to feature in the *History* and the *Philosophy*. Thus he referred to Butler as a 'discoverer' in morals, and suggested that moral ideas derived from a 'law of our nature', but were also formed through interaction with the world, thus anticipating his formulation of the notion of fundamental ideas in the physical sciences (Mackintosh 1836, 40; also Schneewind 1968, 112). In his sermons of November 1837 on the foundation of morals, Whewell explicitly referred to the 'fundamental idea of a moral law', claiming that just 'as in other subjects', its development would only occur when humans set 'their minds upon the reasonings and deductions to which this idea may give rise' (Whewell 1837d, viii–ix). When Hare thanked Whewell for the dedication of these sermons, he agreed that 'the comparison between our scientific and moral knowledge is locally appropriate and very satisfactory' (Hare to Whewell, 4 January 1838, WP, Add. MS. a. 206[171]).

Todhunter regarded 1840 as the point of transition between Whewell's interest in physical science and his later writings on moral philosophy. Fisch sees the *Philosophy* of that year as the conclusion of a study of the nature of excellent physical science, deriving from the early textbooks on mechanics. But neither of these perspectives adequately recognizes the interactions between Whewell's various concerns since the 1820s. When these are considered, the *Philosophy* appears as the culmination of an inquiry which sought to resist secular, utilitarian, and empiricist accounts of ethics, political economy, and physical science. Other writers such as Sedgwick, Jones, Herschel, Rose, and Hare shared these concerns, but Whewell alone produced a systematic and sustained account of inductive science. In doing this he warned against the dangers of a Lockian and utilitarian philosophy of science, one that reduced scientific knowledge to generalizations of empirical observations, just as it reduced ethics to expediency. He did this before such an account was fully elaborated – by Mill in 1843 – but the identification of this erroneous view of science allowed Whewell to defend his idealist epistemology in moral and social, as well as philosophical, terms. Stressing the analogies between scientific and moral knowledge – which Hare appreciated – Whewell presented his epistemology of fundamental ideas as an answer to the threat of moral and social doctrines that rejected the inherent capacities of man's nature and the traditional structure of society (for further details, see Yeo 1979 and Williams

1991). In this sense, the *Philosophy* provided a moral justification of Whewell's role as a metascientist.

Whewell's new sense of his project

During the time in which the new direction of his inductive project was established, Whewell began to present his work as one of moral reform. Before 1832 his dominant *public* concerns were the defence of science against conservative theological criticism and the release of political economy from the control of Ricardian and utilitarian doctrines. In the first case he pursued an argument within natural theology; in the second, together with Jones, he offered both a methodological and moral critique. To some extent, these engagements did connect with his more private inquiry into the nature of the inductive sciences, in so far as the elaboration of a true scientific method could be used to allay religious fears and check the 'scientific' status of political economy. But from 1832 there was a more definite sense of linkage across all three projects, as Whewell came to see that epistemological issues, such as the role of ideal mental conceptions, were involved in both physical and moral science. Furthermore, it is likely that he was able to pursue the idealist philosophy of science which he announced to Jones in 1834, precisely because there was a way of legitimating it as part of a moral and social programme.

This is discernible in the new sense of mission which Whewell was eager to convey to his friends at this time. The announcement of his major project on the inductive sciences was often linked with the debate on ethics. Thus after referring to the contributions of Sedgwick and Birks, he informed Hare that he aimed to write 'a philosophy, such as shall really give a right and wholesome turn to men's minds' (Whewell to Hare, 25 December 1833, in Todhunter 1876, II, 175). And in the course of correspondence with Hare over John Sterling's proposal of a Coleridge Prize for an essay on the 'philosophy of Christianity', he raised the problem of rendering German thought acceptable in England (see Hare to Whewell, October 1834, WP, Add. MS. a. 206[166]).[5] Whewell warned that apart from the likely opposition of those uncomfortable with the unorthodox Christianity

---

[5] This issue of translation had appeared much earlier. In 1825 Hare praised Whewell's mineralogical trip to Freiburgh in 1825: 'I rejoice heartily in the prospect of having their German science so well got up for our market' (Hare to Whewell [28 June 1825] WP, Add. MS. a. 206[154]).

associated with Coleridge's name, the notion of a 'Philosophy of Christianity' had a more definite meaning in Germany than in England. Significantly, he remarked that 'the truths, which may be found in the writings of these men [Germans] must be taken up in the mind of some genuine Englishman and given out in suitable form, before they will take a national hold upon us' (Whewell to Hare, 19 October 1834, in Todhunter 1876, II, 196). Who could this translator be, but Whewell himself?

Two years later Whewell saw a way of delineating a role similar but not identical to Coleridge's notion of the clerisy, one that would legitimate his metascientific project in moral terms. This is reflected in his remark to Rose in 1836 when referring to the claims of Rose and Jones for promotion within the Church: 'I hope your turn will come before long, for I have yet to make out my case by reforming the philosophy of the age, which I am going to set about in reality. I dare say you laugh at such conceit' (Whewell to Rose 12 August 1836, WP, O 15. 47[407]). When Whewell wrote to Hare a year before the Mastership of Trinity became vacant, he said he might leave the university for a Church living because his major work was now done. Hare's reply made it clear that Whewell's character was not suited to the pastoral role of a cleric: he advised that Whewell's vocation was as a doctor, not a pastor, of the Church, saying: 'I would have you pursue your work in moral philosophy' (Hare to Whewell, 17 December 1840, in Stair-Douglas 1881, 211). This certainly gave Whewell a reason for continuing his philosophical work; but if the previous discussion is considered, it is clear that moral questions were already connected with his epistemology.

Apart from offering a response to the secular and radical threat posed by utilitarianism, Whewell also regarded his work as an answer to conservatives within the Church. As Christian Advocate at Cambridge, Rose had complained in sermons of 1827 and 1834 about the arrogant tendencies of 'experimental philosophy' (see ch. 3). Whewell's sermons of 1827 were, as Hare remarked, a 'defence of science from the pulpit' (Hare to Whewell, WP, Add. MS. a. 77[130]). In these, he claimed that 'inductive philosophy' led from observations to laws and finally to 'an Intelligence and a Will from which these laws emanate' (unpublished sermon, 1 February 1827, WP, R 6. 17[14]). Complaints that science encouraged irreligious attitudes stemmed from various errors, such as the failure of some modern philosophers to recall the example of Newton, or from 'something defective and partial in the conduct of their faculties' (4 February 1827, WP, R 6. 17¹³). As discussed in chapter 5, the Bridgewater treatise continued this answer.⁶

But by the late 1830s some of Whewell's friends within the Church were more worried by the influence of the Tractarians than they were by science. Hare wrote to Whewell in 1839, saying that: 'We badly want men of vigour from Cambridge to counteract the modern-antique idolatries of the Oxford school' (Hare to Whewell, 24 September 1839, WP, Add. MS. 206¹⁷⁴). Significantly, when the bishop of London, Charles Blomfield, was asked by Peel about Whewell's qualifications for the Mastership of Trinity, he expressed some doubts about his knowledge of divinity. Was it sufficient for the necessary task of 'checking the extravagances of the Oxford Tractarians'? (Blomfield to Peel, 21 September 1841, in Robson 1967, 315. See Bowden 1839 and Newman 1841 for the critique of science.) In any case, Whewell regarded the High Church and Tractarian attitudes to *science*, rather than their theology, as the target best suited to his abilities. He saw that an answer to their fears about experimental knowledge and scientific institutions could only be successful if the association they perceived between science and the sensationalist philosophy of utilitarianism could be broken. Such a response could not simply affirm the morality of science and its practitioners, as Whewell did in his sermons and Bridgewater treatise; it had to provide an alternative epistemological basis for science.

After the 'Philosophy'

By the time he published the *Philosophy*, Whewell was convinced that it was essential to ensure that physical science was a battleground in the campaign against the utilitarians, and not a form of knowledge they could easily cite as supportive of their epistemology. The Tractarians were happy to abandon science. Whewell was not. But although he confided his strategy, and the sense of moral mission he attached to it, some of his closest colleagues did not fully appreciate his message. In January 1838 Whewell remarked that 'she is rather a

---

⁶ This satisfied Rose: 'you allow what is to be allowed and vindicate science as she deserves' (Rose to Whewell, n.d., WP, Add. MS. a. 211¹⁴³). In a sermon affirming the intellectual value of theology and metaphysics, Rose admitted that 'the most ... which has ever been said for these [physical] sciences, as they can affect the human mind, has been said by one whom I can never name without the strongest emotions of respect and regard' (Rose 1834, 12).

hard task-mistress, this Reforming philosophy of mine; for she carries me into regions where I can hardly expect any of my friends to follow me – at least for a long time' (Whewell to Hare, 7 January 1838, in Todhunter 1876, II, 267). Thus just before the publication of the *Philosophy*, Whewell had to convince Hare – with whom he enjoyed close dialogue – that the epistemological battle lines had to be maintained.

After reading Mill's 'masterly' essay on 'Bentham' of 1838, Hare told Whewell that he was delighted to 'see that party casting away all the worst part of their errors, and incorporating so much of high truth. Everywhere in philosophy it looks as if the extremes are drawing towards each other, and longing to unite in a true and positive centre of union' (Hare to Whewell, 8 October 1838, WP, Add. MS. a. 206[172]). Amazed at the extent of Hare's concession, Whewell insisted that Mill still belonged to the Benthamite party and wrote for its journal, the *London and Westminster Review*, an organ whose professed objectives included the destruction of the Church and the democratization of the nation. Whewell clearly felt somewhat isolated, admitting that he could not understand how Hare's friend, John Sterling, could write in such a journal (Whewell to Hare, 15 October 1838, in Todhunter 1876, II, 270–1). Hare then reprimanded Whewell for being uncharitable to Mill, who was seeking to reject the worst extremes of the Benthamites, even though this meant disavowing the system of his father. Whewell was unrepentant, arguing that the real issue was not the opinions of J. S. Mill but the assessment of an article in the 'London and Westminster Review', a journal 'still supported by one or two of the most thorough-going theoretical and practical destructives and utilitarians'. In frustration he exclaimed: 'So you see I am where I was. I may be ill-natured, but I cannot make out that I am unjust' (Whewell to Hare, 30 October 1838 in Todhunter 1876, II, 272–3).

In the *Philosophy*, Whewell sought to heighten the extremes that Hare saw as fading. His friend's comments were therefore vexing because they implied that he had not appreciated what Whewell was doing: providing the wide 'foundations' for the new system of morals they both desired (Whewell to Hare, 15 October 1838, in Todhunter 1876, II, 270). This anxiety on Whewell's part supports the suggestion that he was trying to vindicate his epistemological position as a contribution to moral and political debates – a strategy that Hare seemed to miss. However, by 1849, when Whewell replied to Mill's

Logic, Hare said that although he had not yet read this book he 'could not understand how any man in these days, with a philosophical head, and a knowledge of what has been done, can maintain the objective origin of all knowledge. At all events I rejoice to find you so zealous and vigorous in maintaining the contrary truth' (Hare to Whewell, 17 June 1849?, WP, Add. MS. a. 206[151]). This meant that Whewell had managed to convince at least one of his closest friends that an anti-empiricist case had to be made in the domain of the physical sciences.

Once Whewell's *Philosophy* appeared, Mill had a clear point of opposition, although his early perception of it invites some comment. In a letter of 5 April 1842 he joked that he seemed to 'stick to the highchurch booksellers. Parker also publishes for Whewell with whom several chapters of my book are a controversy, but Parker very sensibly says he does not care about that' (Mill to R. Fox, 5 April 1842, in Robson and Stillinger 1961–91, XII, 513). This association of Whewell with the High Church is ironic, given his long-term struggle to repel their critique of modern physical science.

Whewell admitted to Herschel that the *Philosophy* was written in a 'spirit of needless pugnacity'; but then said that he knew his 'opinions were opposed to those generally current, and was prompted by that recollection to sharpen my doctrines and my arguments as much as possible' (Whewell to Herschel 26 June [1841] in Todhunter 1876, II, 299). This chapter has argued that Whewell also sharpened the profile of his opponents. As his project shifted from a methodological inquiry within a Baconian framework to an anti-Lockian epistemology, Whewell sought to identify a threat to both science and morals. Once this link was made, his own philosophy of science, with its unconventional epistemology, could be justified as part of a wider moral and social reform. Moreover, Whewell was able to present his metascientific reflections as compatible with his position as a doctor of the Church. However, as the next two chapters show, Whewell's philosophy of knowledge had implications that conflicted with the agendas of other leading men of science.

PART THREE

CHAPTER 8

Science, education, and society

In science, read, by preference, the newest works; in literature, the oldest.

Edward Bulwer-Lytton, in Ebison 1971, 27

OLD AND NEW KNOWLEDGE

In bringing a translation of a German work on university education before the British public in 1843, Francis Newman captured one of the profound tensions of the period. He warned that if 'the new sciences' and all subjects relevant to the physical welfare of society were driven out from the old universities, there would be 'two national minds generated under two hostile systems'. In this contest all intellectual causes would lose and rude 'industrialism' would triumph. Instead, he advised, the ideal should be a curriculum in which the 'moral and material sciences, the modern and the ancient knowledge' grew together, balancing each other (F. Newman 1843, I, xxxiii–xxxiv).

Whewell favoured this notion of balance in his contributions to educational debate from 1837 to 1850, but he did so in a way that left limited space for the new sciences. In the *Principles of English university education* he announced that 'we cannot find in any of the more modern physical sciences, any thing that can fitly be substituted' for the study of geometry and the classics – the traditional course of study at Cambridge. One reason for this was that the definite mental discipline provided by geometry could not be replaced by 'sciences which exhibit a mass of observed facts, and consequently doubtful speculations' (Whewell 1838a, 41). He then mentioned geology.

Some of Whewell's friends within the scientific community believed that he had shipwrecked their cause, placing more obstacles

209

before the proper inclusion of the physical sciences within the undergraduate programme. Charles Lyell, in particular, regarded some of Whewell's views as symptomatic of the forces opposed to the ideal of the pursuit of knowledge in a secular and democratic setting. In his *Travels in North America* – an unlikely place for this topic, as Whewell observed – he reflected on the sorry state of the natural sciences at Oxford in the wake of the Tractarian movement. It was symbolic, he suggested, that the vote of the Oxford Convocation against reforms that might have extended the curriculum, took place in 1839, the year in which the Pope acted against professors – 'of his colleges of Rome and Bologna' – who attended a congress of scientific men assembled at Pisa. He estimated that three-quarters of the sciences nominally taught at Oxford were now 'virtually exiled from the University' (C. Lyell 1845, I, 295–8).

For Lyell, the battle over the inclusion of the natural sciences was not only one between the advocates of a rejuvenated professorial system and the defenders of a tutorial monopoly. It was a contest between those who wanted the universities to serve the widest national goals and those whose interests were narrow, clerical, and elitist. In 1827 he had been hopeful that 'no extensive or violent changes' were required to adapt Oxbridge institutions to the 'wants and the spirit of the age' (C. Lyell 1827, 264, 257; cited in Corsi 1988, 111). But commenting on the response to his book in 1847 he remarked that 'there is a move now in the right direction, but the clerical influence arrayed against all progressive sciences, whether physical or literary, is too powerful to be easily overcome'. In 1850 he told the same correspondent that the clergy were still the 'real educational rulers in this country, which is more parson-ridden than any except Spain' (Lyell to George Ticknor, 2 April 1847 and 1850, in K. M. Lyell 1881, II, 127, 169).[1]

In calling the physical and natural sciences 'progressive' Lyell was ascribing positive intellectual and social value to them. He expected Whewell, as the 'historian of the Inductive Sciences', to endorse this estimate (C. Lyell 1845, I, 304). Lyell's complaint against Whewell was that his foray into the question of university education had created a negative image of the natural sciences. Although agreeing that the learned languages and mathematics should be 'the main

---

[1] Hodge 1991, 259–60 makes the point that Lyell's criticism of Cambridge and Oxford, like Brewster's, may have struck Whewell as an attack on the union of Church and State epitomized in the English, as opposed to the Scottish, universities.

instrument of education', Lyell was surprised that Whewell accepted the monopoly these subjects enjoyed (C. Lyell 1845, I, 303). Replying to Lyell's published criticism in a letter of 1847, Whewell complained: 'I have felt great indignation... You have taunted me with "advocating a monopoly" which I had condemned in the very pages referred to.' The affirmation of science, he protested, had been among his highest priorities, 'on every occasion, public and private' (Whewell to Lyell, 21 March 1847, WP, Add. MS. a. 216[56]). He told Hare that Lyell was bigoted on the university question (Whewell to Hare, 9 May 1848, WP, Add. MS. 215[79]. See Becher 1980, 38–9 and 1986 in defence of Whewell's claims here).

Whewell cited his major works on the history and philosophy of science as testimony to his promotion of these subjects. But Lyell had a point: the opponents of reform could use Whewell's views on education as an argument against the introduction of more science, and would feel no obligation to consult his other works for a broader affirmation of its intellectual and cultural value. In responding to Whewell's letter in a seventeen-page epistle, Lyell explained how he regarded Whewell's educational writings as inconsonant with his status as the historian of the inductive sciences (Lyell to Whewell, 17 April 1847, WP, Add. MS. 216[57]).[2]

However, Whewell did see a definite link between the *History* and the *Principles* – both published in 1837. He predicted that the 'little book' on university education is 'more likely to have a popular currency ... and of course it is founded, actually and professedly, upon the History, as all good books in future should be' (Whewell to Jones, 13 May 1837, in Todhunter 1876, II, 255). On the first page of the *Principles* he distinguished between current debates on university education and his own 'long and somewhat laborious researches on the principles and history of science'. Characteristically, Whewell sought the high ground, claiming that his contribution to contemporary debate derived from a more general, scholarly perspective. Explaining that this wider research project had led him to write the *Principles*, he requested the reader not to 'mix me up in his thoughts' with any local controversies (Whewell 1838a, 1). But there can be no doubt, as Perry Williams has recently argued, that Whewell's educational writings were associated with his critique of utilitarianism and political radicalism (Williams 1991).

---

[2] Augustus De Morgan regretted that Herschel's *Discourse* had not included a strong case for physical science in education. See De Morgan 1832a, 70.

While granting, however, that Whewell did have some similar targets in both the *History* and the *Principles*, it is worth questioning his own claim about the degree of symmetry between them. The difficulty with this account is that it leaves unnoticed some discontinuities between Whewell's historical and educational writings. Whereas the *History* dealt with the themes of revolution and progress, the book on education was pervaded by an affirmation of stability. Although there was, as we saw in chapter 6, a tension around this issue in the *History*, it nevertheless emphasized the importance of novel conceptual leaps in the advance of science. In this work Whewell was writing in a genre largely dominated by an approach that looked forward rather than backward, celebrating the triumphs of the present over the past and looking expectantly to future successes. To some extent, he wrote against this grain, arguing that the historical progress of science was more complex than one in which the present merely shed its past like a shackle. But he did not resist the general message of narrative histories – that progress was the main theme of the drama (Whewell 1857a, 1, 4).

The relationship between old and new knowledge resonated with the subject of university education. But in this sphere, science could not simply be promoted; it had to be assimilated to the traditional ideal of liberal education (Rothblatt 1976; Garland 1980). Thus in the *Principles*, the promotion of the inductive sciences and their contribution to intellectual culture, one of the themes of the *History*, was subordinated to a defence of traditional subjects such as mathematics and the classics. These stressed cultivation rather than discovery of knowledge. Such an attitude jarred with the ethos of the scientific enterprise and its quest for new discoveries. In 1851, Babbage noted that even political conservatives stressed the exponential impulse in scientific knowledge, comparing it with an axiom of political economy: 'accumulated knowledge, like accumulated capital, increases at compound interest' (Babbage, 1851, 211). But research, the engine of this intellectual industry, was not a part of the duties or expectations of the university academic.

This was the point Whewell relied on when responding to criticisms that the English universities were deficient in their attention to modern science. Sir William Hamilton and others had made this charge in articles for the *Edinburgh Review* (Hamilton 1831a and 1831b). Whewell's defence in the *British Critic* referred to the distinction between teaching and research, complaining that it was folly to regard persons in university positions as 'men whose office is *discovery*, or to make demands upon them as if their duty were to produce *new* truths' (Whewell 1831b, 72). As noted in chapter 4, supporters of science such as Thomas Young accepted this fact and looked outside the universities for a diffusion of scientific culture.

There was also a tension between the attitude inculcated in students at university and that encouraged by the scientific enterprise. In university circles in the early nineteenth century, the concepts of novelty and originality still carried negative connotations and were regarded as unnecessary, even dangerous, accessories in the education of young men (Rothblatt 1985, 69). However, in Germany there were calls for the passionate pursuit of new knowledge within the walls of ancient institutions. One of the earliest of these demands came from Friedrich Schelling in his lectures on university studies at Jena in 1803: 'I am fully aware that in the eyes of very many, particularly those who appreciate science only for its practical uses, universities are no more than institutions for the transmitting of knowledge ... in this view, it might seem sheer accident if the teachers, besides communicating knowledge, were also to enrich science with discoveries of their own' (Schelling 1966, 26; also W. V. Farrar 1975, 183).

This was possibly one of the texts Whewell had in mind when he declared that the 'critical spirit' now encouraged in German universities was producing students who flirted with novel philosophies without undergoing the mental discipline offered by the traditional subjects, especially geometry (Whewell 1838a, 45–52). Earlier, in his reply to Connop Thirlwall's pamphlet on the admission of Dissenters to Cambridge degrees and the abolition of compulsory chapel attendance, Whewell had gone beyond the immediate issue to warn of the consequences of loosening regulations. One thing he feared was the notion that 'our University is a place where a man may educate himself, and where the students, by their social intercourse, may enlarge and cultivate each other's minds' (Whewell 1834d, 17). In 1837 he named the German preference for idealist philosophy in the undergraduate curriculum as an illustration of this danger. It had the tendency to set up professors as rival performers in a subject not amenable to clear examination of basic concepts – a feature so well displayed by mathematics and its prominent place in the Senate House examinations (J. Gascoigne 1984, 568). However, this criticism was not clearly distinguished from remarks about chemistry and geology. Whewell's readers, such as Lyell, concluded that both speculative philosophy and the recent natural sciences were being excluded from the undergraduate curriculum because of their novel character.

As the comments of Francis Newman suggest, the argument about liberal education in this period seemed to repeat an earlier controversy: the *querelle* between the ancients and the moderns in the seventeenth century. This involved an attack on the new natural philosophy by defenders of classical models in art, poetry, and philosophy, who dismissed the boasted novelty of recent discoveries as insignificant in comparison with the stable achievements of the Greek and Roman past. Modern science was juvenile and lacked the maturity and stability of ancient knowledge. The champions of 'modern' knowledge, such as Francis Bacon, responded that ancient times were the youth of the world; the present was the true age of antiquity. One compromise was offered by the metaphor of moderns as dwarfs standing on the shoulders of past giants, seeing further because of previous advances (R. F. Jones 1961; Merton 1965). In post-Revolutionary France, the 'moderns' may have won. This at least is what Saint-Simon thought when he said the test of an educated person in 1813 was no longer based on classical learning but on familiarity with the 'positive sciences and those of observation' (Mendelsohn 1964, 9). This was not the case in early-nineteenth-century Britain, where the legacy of the older controversy was manifested in the contrast between classical studies as traditional, and scientific subjects as modern and progressive, in character.

Well before his public involvement in educational debates, Whewell engaged with this rhetoric. His diary of January 1826 refers to an attack on the universities in the *Westminster Review* alleging that while the 'arts and sciences are progressing', the same could not be said of the 'art of education'. Whewell's reply was that the education system had produced this progress (WP, R 18. 9[13]). Eleven years later, in the *Principles*, he developed this point into the claim that modern physical sciences, precisely because they were progressive, should take a secondary position in the university curriculum, even though they had played a positive role in the history of culture. He certainly qualified the fervour of those who asserted that these sciences, because of their recent and continuing discoveries and their practical applications, should either join or displace the study of classical subjects. The twist Whewell gave to the old oppositions was to say that an understanding of modern science, and contributions to their progress, depended on prior mastery of ancient achievements, particularly in geometry (Whewell 1838a).

In doing this, Whewell set some of the dichotomies of the education debate within a general framework – an opposition between 'permanent' and 'progressive' studies or subjects. This was the leading motif of his writing on education and it had implications for his presentation of science.

PERMANENT VERSUS PROGRESSIVE KNOWLEDGE

In the *Principles*, Whewell distinguished between 'speculative' and 'practical' teaching. In the former, he explained, the teacher expounds some branch of knowledge conveying the most recent discoveries, and perhaps his or her own; in the latter, the teacher initiates the student into the foundations of a subject and requires the completion of exercises or problems. In speculative teaching the learner is passive; under the practical method he or she is active. Whewell declared that the practical form of teaching was that of college lectures and tutorials in classics and mathematics, the traditional subjects of the university. The speculative mode was characteristic of professorial lectures in the university and those in the metropolis and provinces on 'physics and metaphysics, geology and political economy, taste and politics'. These subjects could 'hardly be taught otherwise than speculatively' because their 'foundations are constantly undergoing changes'. In contrast, subjects such as geometry and the ancient languages[3] were based on the ideas of space and number and principles of grammar which were 'necessary and immutable parts of the furniture of the human mind'. These subjects thus rested on 'distinct possession of fundamental ideas', whereas the 'wider physical sciences' were only able to present facts as a matter of observation' (Whewell 1838a, 5–9).

Whewell's original distinction between kinds of teaching thus became a qualitative dichotomy between kinds of subjects. He acknowledged this in the *Principles* but did not employ the terminology of 'permanent' and 'progressive' subjects until 1845 in *Of a liberal*

---

[3] Although he regarded classics, like geometry, as part of the permanent studies, Whewell was not enthusiastic about the way these were taught, regarding the method as a mere test of linguistic skill and not a cultivation of reason. See Winstanley 1940, 216–17.

education (Whewell 1845a, 5). The essential distinction was however present in the earlier work where it applied not only to kinds of study, but to kinds of *sciences*. In this classification the permanent subjects of geometry, arithmetic, and algebra, which rested on the 'fundamental ideas of space, number and quantity', were joined by mathematico-physical sciences derived from other ideas, such as pressure and matter, rigidity and fluidity, velocity and force (Whewell 1838a, 9). Thus Whewell was pleased to record that portions of mechanics and hydrostatics had been recently inserted in the undergraduate curriculum. In keeping with another theme of both books, Whewell argued that all these permanent subjects, mathematical and literary, linked the student with the past. In contrast, the 'modern physical sciences' connected humanity with the future. Unlike classics and mathematics they did not 'constitute the *culture*' of the student because they were not subjects by which 'habits of thought' were formed, but they comprised an important portion of the 'information' of *educated* persons (Whewell 1838a, 33, 41). Even when he deliberately adopted the terminology of permanent and progressive sciences, Whewell's characterization of the latter was unclear. In the *Principles* he often grouped them with 'philosophy', a term he also used when referring to the teaching in German universities, and suggested that the new physical sciences were affected by a similar instability in their doctrines. Speaking of the succession of German idealist philosophers from Kant to Fichte, Whewell described it as a case of 'revolution after revolution'. Such studies could not implant the 'conviction of the immutable and fixed nature of truth' (1838a, 9).

The focus on the instability of the 'progressive' sciences here was stronger than in the *History*. In the context of the education debate, Whewell wrote of these sciences as if their theories were merely speculations on a collection of factual observations (Whewell 1838a, 9). In his historical work, there was a more definite statement of the close relation between ideas and facts, of the way in which ideas, hypotheses, and imaginative guesses guided observations, a point not grasped by orthodox Baconianism. Here Whewell indicated that facts only appeared as facts when considered from a particular view (Whewell 1857a, 5–7; see ch. 6). He had come close to this position in 1834 in replying to Brewster's charge that his Bridgewater treatise placed natural theology on uncertain foundations by using evidence from the undulatory theory of light (Brewster 1834, 429). Whewell ridiculed this notion of waiting until 'every whisper of controversy has died away', arguing that there was no hard line between hypotheses and established theories (Whewell 1834c, 266). Yet this is just the approach he adopted three years later in the *Principles*, where he opposed the introduction of progressive sciences in the curriculum. Eventually, Whewell conceded, 'the elements of chemistry and natural history shall be fundamental parts of a good education'; but at present he believed their general theoretical statements lacked clarity, rendering them unsuitable for undergraduate instruction by 'practical' teaching. In 1847 Lyell told Whewell that this was an argument for teaching the present curriculum 'for 100 years', and called him a defender of 'things as they are' (Whewell 1838a, 28; Lyell to Whewell, 17 April 1847, WP, Add. MS. 216[57]).

Arising from this, there were inconsistencies in his general comments on the history of the sciences in the two works of 1837. In the *History*, although Whewell argued that certain knowledge, securely anchored in concepts derived from fundamental ideas, had so far been achieved only in the most advanced physical sciences – astronomy, mechanics, and possibly optics – there was a strong sense in which all the inductive sciences shared in a progressive movement dating from the time of the Renaissance and the scientific revolution. Speaking of the development of astronomy after the 'stationary' period of the Middle Ages, he alluded to the overthrow of obedience to the authority of ancient and scholastic texts: 'The causes which produced the inertness and blindness of the stationary period of human knowledge, began at last to yield to the influence of the principles which tended to progression' (Whewell 1857a, 1, 271–3). Here the spirit of the physical sciences and that of speculative thought seem to be mutually supportive. But in the *Principles* the permanent physical sciences were detached from this spirit of philosophical speculation: instead he stressed that they depended on permanent studies of geometry. Here Whewell postulated that science does not flourish in periods of speculation but in those dominated by the practical teaching of fundamental mathematical concepts (Whewell 1838a, 18–22). In contrast, the later work on *Liberal Education* presented astronomy, optics, and harmonics as progressive subjects developed as part of Greek philosophical activity (Whewell 1845a, 15–16). This inconsistency puzzled Todhunter, who noted that Whewell seemed more favourably disposed towards modern progressive studies in his second book on education (Todhunter 1876, 1, 157).

THE QUESTION OF MATHEMATICS AND THE 'PERMANENT' SCIENCES

It is fairly clear that the tension here was created by the apologetic requirements of Whewell's educational writing. What was he trying to achieve? The positive aspect of his position was a case for the introduction of *physical* science into the curriculum. In *Thoughts on the study of mathematics as part of a liberal education* of 1835, he recommended the inclusion of portions of mechanics and hydrostatics as 'part of degree requirements'. These were examples of that 'great system of physical knowledge which has been steadily advancing' since the Middle Ages. But Whewell was concerned about the kind of mathematics teaching in this curriculum. It was essential, in his opinion, that when these physical sciences were treated mathematically students should grasp the basic physical realities involved. For example, they had to know that 'the conception of *force* and *pressure* is the groundwork of the whole doctrine of mechanics' (reprinted in Whewell 1838a, 156, 174–5). The debate in which Whewell was engaged here predated the more general arguments of the *Principles*; it was associated with his early textbooks, but it intersected with his other educational writing, with some confusing results (see Becher 1980; Fisch 1991a, 36–56; J. Richards 1988, 18–29, for detailed accounts of his views on mathematics teaching).

We should not be misled by the response of Sir William Hamilton to Whewell's pamphlet of 1835. In attacking this in the *Edinburgh Review*, Hamilton constructed a debate between logic – or philosophy – and mathematics as vehicles of the intellectual discipline expected in university education. Although Whewell did refer to this choice, it was, as he indicated in a letter of 1836 to the editor of the review, mentioned only in the first seven pages of the essay (reprinted in Whewell 1838a, 186–9). He was certainly prepared to back the priority of mathematics, but in Cambridge this had been a question settled in the previous century (J. Gascoigne 1984, 570–1 and 1990, 222–5, 246). Whewell's main argument was not with logic but with recent views on mathematics teaching. His quarrel was with those, such as Peacock, Herschel, and Babbage, who wanted to bring Continental analysis into the undergraduate curriculum. As Becher and Fisch have shown, Whewell's concerns about the teaching of such advanced mathematics were apparent from about 1818, and by the 1830s his textbooks contained less calculus than the earlier editions

(Becher 1980, 16–26; Fisch 1991a, 36–42; also 46–8 for the point, against Becher, that such reservations applied only to the teaching of analysis, not to its role in science).

In 1835 Whewell said that mathematics could only be a 'means of forming logical habits better than logic itself' if certain recent errors were avoided. The most worrying was the tendency of those who promoted analysis to be carried away with its elegant form to the detriment of pedagogic need. These mathematicians overlooked the 'peculiar fundamental principles of several portions of mathematical science, substituting for them mere verbal definitions' (reprinted in Whewell 1838a, 163, 166). It is ironic that on this last point Hamilton agreed with Whewell and defended the Scottish approach to the teaching of mathematics against what he saw as the Cambridge influence being exerted via Forbes, Whewell's protégé.

When the Chair of Mathematics at Edinburgh became vacant in 1838, Hamilton wrote to the Lord Provost defending the Scottish tradition. Hamilton protested that his professional interest in 'the cause of liberal education' led him to 'regard *Analysis*, if introduced ... as a substitute for geometry, as an evil'. He warned about losing the solid pedagogic utility of geometry:

The mathematical process in the symbolical method [algebraic] is like running a rail-road through a tunnelled mountain; that in the ostensive [geometrical] like crossing the mountain on foot. The former carries us, by a short and easy transit, to our destined point, but in miasma, darkness and torpidity, whereas the latter allows us to reach it only after time and trouble, but ... while we inhale health in the pleasant breeze, and gather new strength at every effort we put forth. (Cited in Davie 1964, 122, 127)

Whewell may well have been taken by this metaphor; he certainly came close to repeating it in 1845, suggesting that in analysis the mind was 'carried along as in a rail-road carriage, entering it at one station, and coming out at another, without having any choice in our progress in the intermediate space' (Whewell 1845a, 41; Schaffer 1991, 205).

Whewell believed that the inclusion of portions of the mathematico-physical sciences would only be pedagogically effective if they inculcated fundamental conceptions, as geometry did. Consequently, in the works of 1837 and 1845, he deployed the contrast between 'permanent' and 'progressive' subjects in a way that labelled analytical mathematics as 'progressive'. It was necessary for professional mathematicians but not suitable for the basic undergraduate curriculum, which had to ground students in Euclidean geometry.

But this produced a significant confusion. How could analytical mathematics be grouped with the 'progressive' sciences – geology, physiology, and others? Whewell seemed to be saying that, like them, it was still developing, having made great progress 'in the last three hundred years' (Whewell 1845a, 66). But his classification of these subjects as 'progressive' referred to their lack of clear fundamental ideas, or to inadequately stable conceptions derived from them. How could this apply to analytical mathematics, which was elaborating on the fundamental ideas of the most mature physical sciences – Newtonian mechanics, astronomy, hydrostatics, and optics? Whewell seemed to be left with the paradox of considering Lagrangian mechanics as a 'progressive' science in the same sense as geology or biology – sciences which, in his view, lacked the stability of astronomy and mechanics.

Having pushed analytical mathematics outside the standard curriculum, Whewell had to show that part of the physical sciences could be seen as permanent studies, equivalent to geometry and the classical languages. Inevitably, this led him to Newton, especially to the mechanics employed in the *Principia*, which Whewell cited as an example of synthetic geometry in action. When he was enthusiastic about the analytical reforms, he had suggested to Herschel that it would not be surprising if 'in a short time we were only to read a few propositions of Newton, as a matter of curiosity' (Whewell to Herschel, 1 November 1818, in Todhunter 1876, II, 30; Becher 1980, 15). He added that this would sound like treason to most people. But at least by 1832, in *An introduction to dynamics*, which taught the first three sections of the *Principia*, Whewell stressed the value of Newton's 'synthetical proofs' (Whewell 1832d, xii–xiii). In the same year he announced his campaign against those 'strenuous analysts' who saw Newton's works 'in their original form' as a 'millstone about the neck of our system of mathematical education'. On the contrary, Whewell asserted the pedagogic value of 'geometrical and synthetic' methods (Whewell 1832e, vi).

Newton's use of the geometrical method assumed mythical proportion in the *History*:

The ponderous instrument of synthesis, so effective in his hands, has never since been grasped by one who could use it for such purposes; and we gaze at it with the admiring curiosity, as on some gigantic implement of war, which stands idle among the memorials of ancient days, and makes us wonder what manner of man he was who could wield as a weapon what we can hardly lift as a burden. (Whewell 1857a, II, 128)

In his educational works he urged that this method must be studied, not only for pedagogic reasons, but for historical ones. He had mentioned this point in 1832 (Whewell 1832d, xii), but put it more strongly in the textbook on the *Principia* of 1846, where he resisted the erosion of the historical text by successive 'modernization' (Whewell 1846c, ii–iii). This was a classic text, akin to those of Homer or Virgil, which must be studied because it showed how the scientific achievements of the seventeenth century were made. In *Liberal education* he claimed that:

If our knowledge of the Mechanics of the Universe ... do not explicitly exhibit the Newtonian proofs and methods, it will not enable us to share in the interest with which those who know the history of science dwell upon the philosophical events and revolutions of the great Newtonian epoch. (Whewell 1845a, 58)

By thus linking the teaching of selected mathematico-physical sciences to the 'geometrical form' in which they were originally produced, Whewell affirmed their 'permanent' status.[4] Their use of geometrical method was analogous to ancient Greek and Latin: it was a language no longer fully used, but still not dead, since it was the active foundation of modern training and advanced scientific practice. And like the classical languages, 'permanent' science was part of the Western cultural heritage and crucial to good general education (Whewell 1845a, 107).

The previous discussion shows that despite its apologetic value in the education debates, the dichotomy of permanent and progressive studies produced some contradictions in Whewell's more general picture of science. Whewell departed from the usual contrast between ancient and modern knowledge, that is, between classical literary studies and the new experimental philosophy. Breaking with its chronological basis, he grouped the mature physical sciences with the ancient classical subjects. Permanent subjects connected humanity with the past; but in the case of physical science this referred to a period no more than two hundred years ago, since this was when the idea of force was clarified, although it had been contemplated by the Greeks (Whewell 1838a, 27). Thus Whewell's 'long-established demonstrated sciences' were in fact the 'new' sciences advocated by the 'moderns' in the seventeenth century. Indeed in *Liberal education*

---

[4] This may have prescinded the option of conceiving Newtonian science as 'progressive' but still suitable for liberal education.

he acknowledged that the 'discoveries of Newton are still so recent . . . that they may be considered as belonging to the Progressive Sciences; as exemplifying a mental activity which is still going on' (Whewell 1845a, 5, 34–5). Yet in the same passage he argued that Newton's 'exposition of his system has been so long before the world . . . that it may now be very properly used also as a portion of our Permanent Educational Studies' (1845a, 35).

THE QUESTION OF TWO NATIONAL MINDS

Whewell did not explicitly use Francis Newman's idea of a tension in the national culture between the proponents of 'old' and 'new' knowledge. But the problems he addressed were linked with this issue. In the debates about university education this was expressed in the question of how to accommodate scientific knowledge in the traditional curriculum. As we have seen so far, Whewell's response was to reconceptualize *part* of the sciences as already compatible with these values, rather than to challenge the existing rhetoric. This was not achieved without difficulty, or even contradiction, but it allowed the mature Newtonian sciences to join geometry and the classics as part of the 'permanent' curriculum. This was the one that Whewell believed should shape a single national culture; it dictated the terms on which any more recent scientific or other subjects could be included.

A perspective on Whewell's strategy is offered by the comparison with Coleridge's *On the constitution of church and state* of 1830 (Williams 1991, 129). This work gave educational matters a large place in its commentary on English national culture, which Coleridge saw as constituted by a dialogue between permanence and progression. 'In every civilized country,' he wrote, 'the two antagonist powers or opposite interests of the state, under which all other state interests are comprised, are those of PERMANENCE and of PROGRESSION' (Coleridge 1972, 16). His point was that a healthy national culture required a balance between these forces, which he called 'opposite', rather than 'contrary', insisting that, like electrical polarities, they tend towards union and equipoise. There is a definite accord here with Whewell's belief that intellectual advance needed a balance between new and old knowledge, that education must sponsor a dialogue between past and present. But the contrasts here, as well as the similarities, are instructive. We need to recognize that, for Coleridge, these terms represented social forces – such as landed and commercial interests – not academic subjects.

Coleridge promoted the concept of a '*clerisy*', an intellectual class designated as the custodian of the intellectual and cultural pursuits of the nation (1972, 36). Morrell and Thackray have suggested that this concept appealed to the leaders of the British Association in their efforts to elevate the status of physical science. Indeed, there were grounds for this attraction, because Coleridge included the physical sciences in the range of fields from which the clerisy would be drawn, and over which it would superintend. This clerisy comprised the

sages and professors of the law and jurisprudence; of medicine and physiology; of music; of military and civil architecture; of the physical sciences; with the mathematical as the common *organ* of the preceding; in short, all the so called liberal arts and sciences. (Coleridge 1972, 36)

Furthermore, Coleridge attributed advances in the physical sciences and useful arts to the forces of 'Progression'. In his view it was the 'four classes of the mercantile, the manufacturing, the distributive, and the professional' that stimulated progress of 'the arts and comforts of life' and the diffusion of 'information and knowledge' (Coleridge 1972, 17). There was, however, no support here for Whewell's classification of *subjects* as either permanent or progressive.

Did Whewell's distinction between 'permanent' and 'progressive' subjects break this link between science and progressive social forces? His preoccupations during the 1830s may have made this separation attractive. Jones and Whewell were concerned about the way in which the method and success of inductive sciences was being misused by Ricardians, radicals, and utilitarians (see chs. 4 and 7). Whewell had sound strategic reasons for detaching science from the progressivist rhetoric of the reformers and their vision of society based on supposedly scientific principles. First, he wanted to undermine any authority their social and ethical doctrines might draw from physical science and its inductive method. Second, he hoped to place some scientific subjects in the undergraduate curriculum. Both these aims were furthered by the manner in which he removed the most successful physical sciences from 'progressive' studies, leaving this category as a mixture of disreputable German philosophy and new, insecure natural sciences. The latter were definitely important, but their day was yet to come and they still depended on the permanent studies, in which Whewell now included the mathematico-physical sciences. This made it possible to align this last group of sciences with the values he claimed for the 'practical' teaching of stable bodies of

knowledge calling for discipline but not originality. This avoided any
clash with the assumptions of liberal education. It also firmly
distinguished these most prestigious sciences from any involvement in
the utilitarian programme of useful knowledge and social reform. In
other words, it severed any possible link between the best of science
and Coleridge's Progressive forces.

Notions such as intellectual novelty, utility, secularism, and
challenges to authority were associated with the natural sciences by
Lyell, and indeed, by Coleridge. This did not worry Coleridge
because he aimed at bringing all educated groups within a national
clerisy, thereby avoiding the danger of having a disaffected intellectual class – a lesson taught by the French Revolution (Kent 1978, 4;
Allen 1985). But Whewell recognized that any association with
radical doctrines would weaken the place of all science in the
university, and he feared the implications of these notions for wider
debates about the cultural status of science. One result was that
Whewell broke with the form of apologetics conducted on behalf of
science since the seventeenth century, emphasizing not its novelty
and daring speculation, but the possession of stable truths in *some* of its
departments.

The discussion so far has focused on Whewell's contribution to the
debate on university education, one in which he has been variously
typecast as anything from reactionary to conservative reformer. I
have tried to show the difficulties he faced, the arguments adopted,
and the way they related to his other work. Once outside the strict
university context, however, Whewell seemed to be more flexible.

SCIENCE, TECHNOLOGY, AND SOCIETY

In November 1851 Whewell gave one of a series of lectures on the
significance of the Great Exhibition. He opened by observing that he
was 'one of the persons who have the least right of any to address an
audience like this', meaning that he had no part in organizing the
event and had been 'a mere spectator'. But given his silence on the
subject of the practical arts, or technology, in previous writings, this
modesty seems appropriate. In what followed, Whewell gave one of
his most extraordinary performances. Speaking at times in a style
resembling the prophetic declamations of Carlyle, he set the Exhibition in a grand historical panorama, visualizing it as a display that
collapsed the time and space between nations, thus allowing a

comparative perspective on human civilization. It was the grandest event of this kind since the vast body of arts and inventions, stemming from the Middle Ages, accumulated in the sixteenth century – the 'last great Exhibition' (Whewell 1851a, 1, 4, 6; on the comparative vision here, see Stocking 1987, 1–6). Was this a change of mind? How could Whewell praise recent technological advances without lending support to the view that the value of science lay in its utility?

It is instructive that Whewell used the image of a railroad when castigating the negative, mechanical tendency of modern analytical mathematics, for prior to his speech at the Great Exhibition he was not noted for his positive statements on technology. For example, when he translated Brewster's pleas for a 'direct national provision for men of science' into a worst-case scenario it was, as he told Forbes, one in which 'you and I should tell people where to establish fisheries and railroads!' (Whewell to Forbes, 14 February 1835, in Todhunter 1876, II, 204). As noted briefly in chapter 6, technology did not feature in the story of human intellectual progress outlined in the *History*. Brewster's attack on this work made much of its inadequate treatment of the practical arts and, in particular, the links with the scientific achievements of Scots such as Watt, Black, Robison, and Leslie (Brewster 1837, 146–8; also Davie 1964, 175–9). Whewell's response was that Brewster's 'uproar about steam-boats and gas and railways shows that he has not at all comprehended the nature of the book' (Whewell to Jones, October 1837, in Todhunter 1876, II, 261). But there was more involved here than a simple case of selection, because Whewell opposed what he regarded as the exaggerated contemporary adulation of technology. The *History* was in part an attempt to ensure that science, and not just technology, was adequately celebrated in the early Victorian age (see diary entry for 1827 cited in ch. 6; and Whewell 1833a, xxiv–xxv, for his first public statement on this).

The topic of practical inventions made only one appearance in this work, in the section on the Middle Ages, a period Whewell extended to 1500. Here he was anxious to ensure that the practical achievements such as printing, gunpowder, clocks, and telescopes did not count as scientific advances. What general principles in the relevant sciences, he asked, did these inventions disclose? Acknowledging that they may have changed the world, he maintained that 'in the history of the principles of the sciences to which they belong, they may be omitted without being missed' (Whewell 1857a, II, 252–5). Here the distinction between art as practical and science as speculative came into play. Art was 'the parent, not the progeny, of Science', in that it produced examples of processes often later explained theoretically by science. Significantly, Whewell applied this dichotomy to more recent technological advances which, in his opinion, were wrongly ascribed to science: he wanted science to be lauded justly for its intellectual value, rather than incorrectly for its useful application.

Whewell returned to this topic in the *Philosophy*, explaining why the arts were not included in his classification of the sciences, just as they were not treated in the *History* (Whewell 1840a, i, xiii–xiv). Extending the point about the practical, rather than theoretical, nature of the arts, he suggested that 'the principles which Art *involves*, Science alone *evolves*' (Whewell 1847a, ii, 111). And when he discussed the French encyclopaedists as followers of Locke, one of the underlying reasons for this attitude became apparent. He remarked on the way in which Diderot and D'Alembert placed the mechanical arts 'side by side' with the sciences in their publication. They also regarded the greater prestige of 'the liberal over the mechanical Arts' as a prejudice, esteeming inventions as at least equal in value to science, and measuring the value of science 'by its utility' (1847a, ii, 303–6). This all fitted with their sensationalist epistemology: not recognizing the crucial role of ideas, they did not appreciate the difference between a process and its theoretical explanation.

With some justification, these comments have gained Whewell the reputation of being indifferent to the practical applications of science, or at least with ensuring that this issue was marginalized in the British Association (Sanderson 1972, 4; Morrell and Thackray 1981, 256–60). But it is worth observing that Whewell's attitude was not unusual. For example, Lyell and others who complained about his inadequate support for the educational claims of the physical sciences raised no protest against his characterization of the distinction between science and technology. Indeed, Lyell protested that 'if the practical connection of any branch of science be not obvious ... scarcely any encouragement is given to it' (C. Lyell 1845, i, 310). Even later, Lyon Playfair, who promoted the industrial application of science, endorsed the priority of abstract science, as Whewell did (Yeo 1981, 78).

It is also revealing that Macaulay and others felt that Brewster had broken convention in his reviews of Whewell's works by introducing political and educational issues, such as the conflict between

Cambridge and Edinburgh. Macaulay's point was that Whewell was treating 'a question of pure science', and thus one that deserved to be discussed on its own merits (Macaulay to Napier, 24 January 1842, in Napier 1879, 361; Davie 1964, 178). But Brewster was asking about the social context and application of science. Macaulay's complaint effectively kept this outside the philosophy of science – precisely where Whewell wanted it to be. Whewell was adamant that the 'application' of scientific theories was quite distinct from the nature of science – his topic – which was 'purely intellectual':

The augmentation of human power and convenience may impel or reward the physical philosopher; but the processes by which man's repast are rendered more delicious, his journeys more rapid, his weapons more terrible, are not, therefore, Science. (Whewell 1847a, II, 112)

This passage was reprinted, unaltered, from the first edition of the *Philosophy* of 1840. But at the Exhibition in 1851, Whewell seemed to acknowledge a new field for speculation. The objects on display represented '*material* art', and these, just as the other arts of poetry and painting, deserved their 'Critic'. In the first ages of the world, he explained, Tubal-Cain was 'the instructor of every artificer in brass and iron; but it was very long before there came an instructor to teach what was the philosophical import of the artificer's practices' (Whewell 1851a, 13). Showing that he was acquainted with existing analyses of technology, Whewell spoke with authority on the various ways in which the objects had been classified at French Expositions since 1806. A generally accepted classification of material and instruments, he suggested, would mean that the 'man of science, the artisan, the merchant, would have a settled common language' (1851a, 8–12). Babbage raised this point in *On the economy of machinery and manufactures* of 1832, where he predicted a closer connection between the practical arts and the 'severer', abstract sciences, linking this with an observation about the changing social background of scientists:

It is highly probable that in the next generation, the race of scientific men in England will spring from a class of persons altogether different from that which has hitherto scantily supplied them. (Babbage 1835, 383–4)

It is worth recalling that Whewell came from the social class to which Babbage looked for future men of science. Indeed, as the son of a Lancaster carpenter (as Morrell has recently noted), Whewell had earlier and closer acquaintance than Babbage with practical machinery – winches, lathes, pulleys – an inheritance manifested in his invention of an anemometer in 1834 to measure the direction and velocity of the wind (Morrell 1992, 111). And in spite of the omission of technology from his major works, he did favour the use of examples from engineering in the mathematics curriculum. Writing to Babbage in 1832 he professed the intention 'to make Cambridge Mechanics bear upon actual practice' (Whewell to Babbage, 30 April[?] 1832, Babbage papers, BL Add. MS. 37187 f. 196). To some extent he realized this in *The mechanics of engineering* of 1841, which he hoped would bring a practical study into contact with explicit theoretical principles. Reciprocally, he suggested that:

if the common Problems of Engineering were to form part of our general teaching in Mechanics, this science also might become a permanent possession of liberally educated minds... And this would, I conceive, be an improvement, not only in professional, but in general education. (Whewell 1841b, vi)

By 1851 Whewell was no longer silent about technology. He spoke knowingly and with admiration of manual and mechanical skills and inventions. Indeed, he proposed that 'such a great spectacle of the works of material art ought to carry with it its scientific moral' (Whewell 1851, 4). In beginning to unfold this he even seemed to dilute the distinction between art and science that characterized his earlier writings, now describing 'science' as the effort to discover 'the laws of operative power in material productions, whether formed by man or brought into being by Nature herself' (1851, 3). Another observation stimulated by the Exhibition was that recent developments in the chemical industries suggested a change in the relationship between science and art. In this field, rather than merely explaining the principles behind existing technology, science was 'the whole foundation, the entire creator of the art'. In this case, 'Art is the daughter of Science' (1851, 13). Now science could be expected to produce technological applications, rather than to follow practical artefacts. This was significant, in Whewell's opinion, because it offered a definite illustration of how knowledge was power, a point not evident when Bacon prematurely made this declaration. Elaborating on this later, he listed the steam engine, railway engineering, and the electric telegraph as modern examples involving 'recent and profound discoveries in theoretical science' (Whewell 1857a, 306–8). In the *Novum organon renovatum*, part of the third edition of the

Philosophy, he included additions on 'the Application of Science', admitting that a proper treatment would 'require an acquaintance with practical arts and manufactures of the most exact and extensive kind'. Nevertheless, the key feature already visible was the closer 'union of Art with Science' in recent times (Whewell 1858a, x; compare Babbage 1851, 19–20, 131).

For Whewell, however, none of this meant that the line between science and technology was blurred. It meant that metascientific commentary should include these new developments, showing, for example, that in some cases science now led rather than followed art. But science could not be judged by its applications and these could form no part of the impulse which directed it. At least in print, he did not go so far as Herschel, who raised the question of the involvement of science in social and political action. Inspired by the grand themes of Alexander von Humboldt's *Kosmos* and by the events of 1848 in Europe, Herschel referred to the attempts to avoid war and increase the happiness of growing populations. Yet since the success of this political effort demanded that nature be exploited for 'human wants', the progress of science was crucial: 'Science must wave her magic wand.' At this point he held back, perhaps recognizing that such a participation in political goals would threaten the independence of science – its claim to 'free, undisturbed, and dispassionate thought' (Herschel 1848, 182–3; Ross 1978, 80–1). For Herschel, as noted earlier, this meant the untrammelled intellectual freedom of individuals.

Whewell had no problem with the notion, expressed by Herschel, that the arts and sciences could contribute to social and moral progress. Indeed, in asking whether the arts and manufactures of the modern European nations were superior to the extraordinary technical achievements of Persia, India, and other 'Oriental' civilizations, he concluded that the difference was political rather than intellectual. If technical skill and complexity were the criteria, it was difficult to describe the modern European arts as more 'advanced' than those of previous societies. But in social terms, Western nations were more 'advanced' because there was a diffusion of technology among the whole of a population. There was a difference, then, between those countries in which 'the arts are mainly exercised to gratify the tastes of the few' and those in which they supplied 'the wants of the many' (Whewell 1851a, 8).

One assumption here was that art (or technology) did not display intellectual progress in the way that science did; it was only possible to speak of measurable improvements in this sphere when the laws and principles underlying particular arts, processes, or inventions were disclosed – that is, by advances in *science*. Thus Whewell's original position on this issue remained fundamentally unaltered, in spite of his enthusiastic and informed analysis of the Great Exhibition. When he next lectured general audiences on the role of science in education he cast it as the third great intellectual element of Western culture, succeeding the geometry of classical Greece and the jurisprudence of ancient Rome. The achievements of the physical sciences since the seventeenth century, like those in geometry and law, could all be called 'sciences' in the older sense of *scientia* (Whewell 1854a, 4, 16). There was no reference to technology in this lecture, because unlike these other subjects it could not be considered as an intellectual discipline suitable for the cultivation of minds, even though its effects might enhance the mode of life. Significantly, however, the 'progress-ive' physical sciences, and their history, were now firmly presented as suitable examples of stable truth in this general education (Whewell 1854b).

CHAPTER 9

The unity of science

I have been once to the top of Snowden but as you will find if ever you take to mountain climbing the most picturesque part of a mountain is by no means its top.

Whewell to his sister, 9 September 1818, WP, Add. MS. a. 301[2]

At the time of the foundation of the British Association there was anxiety about the fragmentation of the 'commonwealth of science'. Whewell coined the term 'scientist' in this context; and in spite of the fact that Faraday and others had begun to demonstrate the connections between various physical forces, these concerns continued. As President in 1859, Prince Albert feared that specialization would weaken the 'consciousness of its unity which must pervade the whole of Science' (Albert, Prince Consort 1859, lxiii). Later commentators such as Todhunter and James Clerk Maxwell regarded Whewell's work as a counterweight to this tendency:

He [Whewell] did not despair of the fortunes of knowledge; he held that real connexions must subsist among the apparent infinite mass of details, and he encouraged investigators to seek for such general ideas in all their occupations. The lessons which he explicitly taught and implicitly suggested seem peculiarly useful at the present day when every pursuit is specialized. (Todhunter 1876, I, 112; also Maxwell 1876, 206)

Possibly as another sign of resistance, the intellectual landscape of the nineteenth century was marked by some large synthetic systems. The most prominent were those of the German idealists, Hegel and Schelling, but in France Victor Cousin and Comte also offered systems that were universal in aspiration. The works of Bentham, Coleridge, and Spencer were probably the closest counterparts in England, while in Scotland Reid, Stewart, and Hamilton also covered a large number of topics, although most of these reflected the curriculum they taught. Whewell, however, did not attempt a systematic account of the standard repertoire of philosophical subjects ranging from logic to ontology. Compared with these writers, Whewell's focus was more deliberately on natural science.

But Whewell's interest was not confined to the physical sciences: he aspired to a general philosophy of knowledge. In the *History* he stressed that its concentration on the natural sciences did not imply that those subjects not discussed, such as 'Ethology and Glossology, Political Economy, Psychology', were outside the 'rank of Inductive Sciences' (Whewell 1857a, I, 15). Similarly, in the *Philosophy*, he said that although subjects most commonly termed 'sciences' were those dealing with the material world, they could be studied for 'lessons' of wider applicability:

The views respecting the nature and progress of knowledge ... though derived from those portions of human knowledge which are more peculiarly and technically termed Sciences, will by no means be confined, in their bearing, to the domain of such Sciences as deal with the material world, nor even to the whole range of Sciences now existing. On the contrary, we shall be led to believe that the nature of truth is in all subjects the same, and that its discovery involves, in all cases, the like conditions. (Whewell 1847a, I, 3)

Whewell thus hoped to uncover the common features of a single philosophy of discovery, to show that 'the progress of moral, and political, and philological, and other knowledge, is governed by the same laws as that of physical science' (1847a, I, 7).

Furthermore, as we saw in chapter 7, the idea of a relation between the philosophy of the physical sciences and other moral issues was central to the rationale of Whewell's metascientific vocation. He contended that an erroneous theory of scientific knowledge, such as that of the sensationalists and utilitarians, sponsored dangerous moral and social ideas. As a safeguard against such errors, Whewell proposed that 'a just Philosophy of the Sciences may throw light upon the nature and extent of our knowledge in every department of human speculation' (Whewell 1847a, I, 3).

When he wrote the *Philosophy* in 1840, Whewell presented it as the basis of a philosophy of knowledge that would embrace the moral and social sciences. But the stakes here rapidly escalated. As mentioned in chapter 4, Chambers in his *Vestiges* of 1844 showed how natural science could be used as the starting-point of a sweeping system, leaving Whewell dismayed at the way in which the nebular hypothesis could be exploited as the basis of a social philosophy (Yeo 1984, 23; Schaffer 1989, 147, 155). From the late 1830s, Comte and his English disciples made the move from physical to social and political philosophy at breath-taking speed. Harriet Martineau's translation of Comte in 1853 gives some idea of the schedule envisaged:

While our sciences are split up into arbitrary divisions; while abstract and concrete sciences are confounded together, and even mixed up with their application to the arts, and with natural history; and while the researches of the scientific world are presented as mere accretions to a heterogeneous mass of facts, there can be no hope of a scientific progress which shall satisfy and benefit those large classes of students whose business it is, not to explore, but to receive. (H. Martineau 1875, i, vii)

Of course, Whewell acknowledged the need for classification of science that Comte prescribed in his *Positive philosophy*, but Martineau's call was a continuation of Chambers's message. Whewell could not accept the broader programme that was now attached to the analysis of science he had outlined in 1837 and 1840. The 'growth of scientific taste among the working classes', which Martineau hailed as one of the 'most striking of the signs of the times', was at most a mixed one for Whewell, precisely because of the way in which writers like Chambers used science (H. Martineau 1875, I, vii).

Comte's work and the writings of his followers represented a synthetic system founded on an account of science. To some extent this meant that after 1840 Whewell's metascientific project was faced with a competitor that inflated all expectations. Although Whewell aimed the *Philosophy* at 'metaphysical readers', he apologized to some friends for its length and abstraction: 'you will be alarmed at my two volumes', he warned James Spedding (Whewell to Spedding, 17 November 1848 in Stair-Douglas 1881, 356). But he did insist, in a reply to a review by De Morgan, that he had taken care to 'reject, for the present, all that was manifestly vague, obscure, and doubtful' (Whewell 1840b, 4). Whewell certainly believed that the book had a message for the wider public, but acknowledged that it would not be a popular success. It is significant therefore that when, at the end of his life, he finally responded to Comte, in his last review article, Whewell admitted that the 'Positive Philosophy' was a topic of 'popular interest' (Whewell 1866, 353–62; see Whewell to Jones, 14 July 1839, in Todhunter 1876, II, 281 on 'reception' of *Philosophy*).

Although Whewell managed to postpone a specific response to Comte until the end of his life, his work was linked with Comte's by Brewster in 1838. In outlining different modes of involvement in science, Brewster recognized the metascientific role of persons who, rather than doing substantial research, sought to become 'the legislators of science'. There were only two modern writers who fitted this role: Comte and Whewell. However, Brewster decided that Whewell was the less prestigious of these two legislators. Why, he asked, did Whewell's *History* (which he had reviewed the previous year) make no reference to the *Cours de philosophie positive?* Had Comte's work not found its way to Cambridge, even though it 'was well known and highly appreciated in London' before the appearance of Whewell's book? (Brewster 1838a, 273–4.) Furthermore, when he reviewed the *Philosophy* in 1842 Brewster deployed Comtian language in dismissing Whewell's concept of fundamental ideas as 'scholastic Metaphysics' (Brewster 1842a, 266).

Like Whewell, Comte aspired to define science; thus Whewell responded by attacking Comte's credentials at this level. The French writer was 'a person whose want of knowledge and of temperate thought caused his opinions on the philosophy of science to be of no value' (Whewell 1866, 353). It was not too difficult to make such allegations before a scientific audience: Herschel did so at the British Association meeting in 1845, castigating Comte's mathematical treatment of the nebular hypothesis (Herschel 1845, xxxviii–xxxix; also Mill to Herschel, 9 July 1845 and Mill to Comte, 18 July 1845, in Robson and Stillinger 1981–91, XIII, 673–5, 677–9). However, Whewell was conscious of taking this case to a broader audience – the readers of *MacMillan's Magazine* – knowing that Comte had been favourably discussed in the *Fortnightly Review*. Identifying himself as one of the few English readers of Saint Simon (in the early 1830s), he said that Comte had not done justice to the subtlety of these theories of society.[1] Conveniently, he was able to draw on Mill's disillusion with Comte's speculations on sociology and the religion of humanity. Apart from using this to question Comte's sanity, Whewell decreed that this area was outside the philosophy of science proper and did not concern him (Whewell 1866, 359).

But while it was possible for Whewell to dismiss the Comtian synthesis as over stretched and based on a flawed account of science, it was not so easy for him to ignore the demands of those who sought to

---

[1] Jones had claimed 'that Comte is a child of the St. Simonian's without ... their philosophical cleverness' (Jones to Whewell, May 1843, WP, Add. MS. c. 52[81]). Whewell preferred not to discuss Comte's social views because, with Jones, he saw them as a threat to existing institutions. See Jones to Whewell [1843], WP, Add. MS. c. 53[81].

use recent philosophy of science in a new natural theology. After all, this is what Whewell believed his own work allowed, having advertised the point by repackaging selections from his major works as *Indications of the creator*, when asked to reply to the author of *Vestiges*.

As is well known, Darwin was able to use a quotation from Whewell's Bridgewater treatise in the frontispiece of the *Origin* to support the notion that the material world was governed by general laws. But from at least the 1830s other writers were recasting the old Paleyan natural theology on this principle (R. M. Young 1985; Brooke 1991b, 192–213). In some cases, this was expressed as a commitment to the 'principle of uniformity' – the efficacy of a single method of investigation across all the sciences, leading eventually to the discovery of uniform, general laws covering all natural phenomena (for appeals to method as the guarantor of unity, see Yeo 1986a). Babbage anticipated such an outcome in his *Ninth Bridgewater treatise*:

All analogy leads us to infer, and new discoveries continually direct our expectation to the idea, that the most extensive laws to which we have hitherto attained, converge to some few simple and general principles, by which the whole of the material universe is sustained, and from which its infinitely varied phenomena emerge as the necessary consequences. (Babbage 1838, 32)

The most frequent and consistent advocate of this view was Baden Powell. Impressed by the comprehensiveness of Comte's system when he reviewed it in 1839, Powell believed that the Church needed a synthesis of similar breadth in order to remain in touch with the progress of science. But Whewell could not accept the conception of unity advanced by Powell, even though he shared the aim of demonstrating that science, rightly understood, was compatible with Christianity.

Powell began writing on the relations between science and religion in the 1820s, and adopted a fairly liberal approach by the late 1830s (Corsi 1988). He stressed that the advance of science was revealing general, and increasingly comprehensive, laws of nature; these, not miraculous interventions, evinced the existence and power of the Divine Creator. In taking this position, Powell aligned himself with those who held that inductive method rested on '*a conviction of the universal and permanent uniformity of nature*'. On this view, scientific explanations assumed the uniform and universal action of presently known causes and then, legitimately, extrapolated beyond readily observable phenomena (Powell 1849, 35 and 1859, 227–34).

For Powell, then, the unity of science consisted in the adherence to this principle of uniformity. Thus in *The connexion of natural and divine truth* of 1838 he rejected the suggestion that geology, in dealing with unwitnessed past events, should be distinguished in any way from other inductive sciences. Here he clashed with Whewell's contention, in the *History*, that such 'palaeotiological' sciences, because they searched for historical causes, could admit a *qualitatively* different order of past forces that lay beyond the scope of inductive inquiry. Powell argued that this position was tantamount to conservative theological attacks on geology. Moreover, it excluded the discipline from the 'type of unity which binds together the whole range of inductive sciences' (Powell 1838, 60–6 and 1855, 52–9). Later we will see that this example was part of a significant conflict between Whewell's philosophy of science and the strategy of his Oxford counterpart. What features of Whewell's work upset the campaign for a rational natural theology as conceived by Powell and others? To answer this question we need to consider carefully the meaning of the 'unity of science' in Whewell's thought.

ANALOGIES BETWEEN MORAL AND PHYSICAL SCIENCES

When Mill reviewed Whewell's *Lectures on systematic morality* in 1852 he claimed that the 'intuitionist school' relied on two contrasting methods. In physical science they were Baconians; in morals they remained Cartesians. Mill predicted that a dependence on intuitional concepts in ethics would lead to a priori speculation about the physical world, as it had done in the case of German Idealist philosophy (Mill 1852, 352–3). But as the occupant of the Knightbridge Chair in moral philosophy from 1838 Whewell opposed this outcome, stressing the role of empirical and ideal elements in both science and ethics. By this time, he had published the *History* and was writing the *Philosophy*. In his introductory lectures on moral philosophy, Whewell explained that although these works seemed remote from the duties of the Chair, such inquiries into physical science might soon 'afford a basis for all philosophy, even that of morals, and, in some measure, an example of the mode in which philosophical truth may be sought (Whewell 1841a, 30–1). This programme related back to his earlier discussions with Jones on the case of political economy as a moral science. But now, with the two major works completed, Whewell was more confident about the grounds for an extension of the inquiry into the '*hyperphysical* sciences' – the area he had earlier mentioned to Jones as the third part of his project (Whewell to Jones, 27 July 1834, in Todhunter 1876, I, 90 and 21 August 1834, in II, 187; see ch. 7 above).

Whewell began by confronting the common scepticism about analogies between the moral and physical spheres. He admitted that when speaking of 'Moral and Political Sciences ... we never use the word without a misgiving'. There was a common opposition between physics as the investigation of the material world and ethics as the study of human moral behaviour. But Whewell contended that a focus on human action did not establish 'any essential difference between the nature of truth, or the mode of seeking it, in morals and in physics', for such action was involved in all departments of scientific knowledge. Just as cases of moral action necessarily preceded systematic thought about them, so also in physical science, action, either in the sense of practical art or deliberate experiment, preceded theory. Challenging the view that moral knowledge was 'vague and precarious', while physical knowledge was 'exact and certain', Whewell contended that the latter was not usually perfect and the former not entirely obscure (Whewell 1841a, 34, 42).

Following these lectures Whewell published *Elements of morality* in 1845 and *Lectures on systematic morality* in 1846. In both he took a strong position on the relationship between moral and physical science. In the first he claimed that the basic propositions of a moral system were analogous to 'the *Axioms* in Geometry' (Whewell 1845b, viii; Schneewind 1968 for a detailed account). This was part of his answer to utilitarian ethics: by asserting the possibility of a system of 'Moral Truths, definitely expressed, and arranged according to rational connection', Whewell was counteracting the view that such notions were solely derived from the effects of pleasure and pain, and justified by utility (Whewell 1846b, 25; also Whewell to Herschel 3 July 1846, in Todhunter 1876, II, 337–9). However, even writers sympathetic to his aims were dubious about this shift from quite general analogies between moral and physical science to the specific comparison of morality and geometry. Frederick Myers could not accept any similarity between the elements of morality and 'the elements of Euclid' (Myers to Whewell, 31 July 1845, in Stair-Douglas 1881, 323).

Whewell believed that resistance to such comparisons reflected a mistaken view of knowledge. Morality was often said to depend on facts revealed by consciousness and physical science on facts gained by observation. But this distinction did not support an extreme antithesis:

For in mechanics too, those who treat of the elements of the science, have been in the habit of referring to our consciousness. They have usually taught that we derive the conception of *Force* from the muscular effect which we are conscious of making. And even in geometry, on what evidence do axioms depend? Is it not on the impossibility which we are *conscious* of, when we try to conceive the opposite of them? (Whewell 1846b, 53)

Indeed, in arguing with Myers, Whewell was prepared to extend the analogy between morals and geometry to morals and mechanics. The latter offered a clearer illustration of the progressive character of both scientific and moral knowledge. He told Myers that 'although we now have axioms, defined conceptions and vigorous reasoning [in mechanics], we can point to persons, and to whole ages and nations, who did not assent to those axioms because the mechanical ideas of their minds were not sufficiently unfolded'. Moral knowledge displayed a similar process, and hence writers such as Paley were misguided in testing the intuitive character of moral ideas by asking whether they were evident to all people, including savages. For Whewell, axioms in morality, geometry, and mechanics became evident to individuals only as their conceptions were 'rendered clear and definite'; this required a certain 'culture' of the mind. (Whewell to Myers, 6 September 1845, in Stair-Douglas 1881, 327–8; Whewell 1846b, 34–42). This argument was part of Whewell's attempt to demonstrate that an intuitional theory of ethics, or 'high morality', conformed to his philosophy of the inductive sciences: namely, one that stressed the historical process through which fundamental ideas were gradually clarified (Whewell 1862, 1–3, 126–32).

But Whewell's attempt to affirm the unity of moral and physical knowledge was not without problems. The first danger was that the special character of moral ideas might be undermined. The tendency to assimilate morality too closely to the physical sciences weakened the point he was trying to make: namely, that moral ideas derived from human nature rather than from the consequences of human actions. In reply to this he insisted that the pursuit of analogies did not cancel all differences between moral and physical knowledge. Although there were 'conditions of knowledge' sufficiently general to embrace all truths, moral conceptions had to be 'treated by moral arguments; not defined and reasoned about in a mathematical manner' (Whewell 1846b, 44–5, 48–9).

Second, in attempting to save the analogy between physical and moral science, Whewell stressed differences *within* the physical sciences. He adopted this tactic in his first lectures on moral philosophy:

when we look at such sciences as meteorology, and physiology, and geology, we see that the indefiniteness and obscurity of the subjects contemplated, cannot be a reason of imperative authority to exclude morality from the list of sciences, in the most genuine sense of the term; that is, bodies of exact, systematic, progressive knowledge. (Whewell 1841a, 37; also 38–41)

This was a way of defending the systematic study of morals, but it exposed the considerable range of knowledge within the *physical* sciences. Whewell's discussion with Myers on the relationship between morals, geometry, and mechanics raised the issue of different scientific disciplines. When Myers was prepared to accept a comparison of morality with 'the lower orders of the Inductive Sciences' rather than with geometry, Whewell said that he 'did not know to what classifications of high and low order of sciences you refer' (Whewell to Myers, 6 September 1845, in Stair-Douglas 1881, 328). But Myers may well have been thinking of Whewell's own distinction between 'permanent' and 'progressive' sciences. If Whewell wanted to conceive of morality as a body of 'permanent truths', then this division was pertinent (Whewell 1846a, 62). For example, his view of moral knowledge was not consonant with his definition of the 'progressive' physical sciences as those not yet stabilized by a clear fundamental idea. Nor did it fit comfortably with the notion that the emergence of new fundamental ideas could radically alter the theories of a subject (on this problem see Schneewind 1977, 111–17 and Yeo 1991a, 189–90).

We have seen here that, like many other writers, Whewell found difficulty in maintaining the ideal of a unity across moral and physical sciences. In addition, however, even when restricted to the *physical* sciences, the notion of unity in Whewell's philosophy was a complex one. The conflict with Powell stemmed from this second set of tensions.

LIMITS TO UNITY IN WHEWELL'S PHILOSOPHY OF SCIENCE

Whewell's confidence in the potential relevance of his *History* and *Philosophy* to other fields of inquiry rested on an inference he shared with Comte and Powell. The striking success of similar methods in a range of disciplines, and the growing integration of different areas under fewer theories, excited these thinkers. Whewell observed that as human knowledge of the material world developed, 'certain universal characters' or 'general laws' were revealed. His two major works offered '*a connected and systematic survey of the whole range of Physical Science*' and, reflecting on this, Whewell declared that it was possible to envisage the various sciences as 'all members of one series, and as governed by rules which are the same for all' (Whewell 1847a, 1, 8). Considering the advances in sciences from astronomy to physiology, he recalled Bacon's aspirations and asked whether these developments manifested 'some common process, some common principle'; whether 'the organ by which discoveries had been made had something uniform in its structure and working'. If so, it might be possible to distil from the history of scientific progress 'the elements of a more effectual and substantial Method of Discovery', one applicable to new areas of inquiry. The English Comtians certainly believed this; but Whewell restrained this optimism, warning against the error of forcing analogies between, say, the nature of physiological discoveries and those of mechanics (1847a, 1, vi, 9).

The comprehensiveness of Whewell's survey, which included what he called 'intermediate sciences', alerted him to the pitfalls of such generalizations. He therefore qualified the Baconian concept of a common method capable of simple transfer across various fields: although tempting, an *Art of Discovery* was not strictly possible, partly because historical study suggested that 'different sciences may be expected to advance by different modes of procedure, according to their present condition' (Whewell 1847a, 1, viii). There were various stages in the development of science – chemistry, for example, was in its infancy compared with the maturity of astronomy – and different modes of inquiry were appropriate. As we saw in the discussion of Whewell's *History*, one of the most crucial moments in the development of each branch of science was the grasp of the peculiar fundamental idea that constituted the conceptual framework of a particular discipline. In summarizing this thesis in the *Philosophy*, Whewell emphasized that 'each science has for its basis a different class of Ideas; and the steps which constitute the progress of one science can never be made by employing the Ideas of another kind of science' (1847a, II, 18–19). This insight set limits to Whewell's notion of the unity of science. Although willing to explore links between different branches of science – such as mineralogy and chemistry – Whewell was extremely wary of attempts to explain substantially new phenomena in terms borrowed from a previously established science (Whewell to Herschel, 15 February 1829, in Todhunter 1876, II, 98; Whewell 1834a, 337).

This did not mean that the idea of unity had no force. The recognition of differences between disciplines did not banish the promise of a general unity, because it was legitimate to conceive '*definite* differences' among the sciences as included 'within a finite boundary of similarity' (Whewell 1841a, 32). Thus, despite his celebration of the success of mathematico-physical sciences, Whewell took care to say that less exact subjects did not fall outside the category of 'science'. For example, the imprecise character of classification in natural history did not render the subject unscientific. Judgements of this kind arose from the fact that

the mathematical and mathematico-physical sciences have, in a great degree, determined men's views of the general nature and form of scientific truth; while Natural History has not yet had time or opportunity to exert its due influence upon the current habits of philosophizing. (Whewell 1847, I, 494)[2]

From this perspective, Whewell believed that there could be a movement from 'the Philosophy of the Sciences' to the 'Philosophy of Science' (Whewell 1847, I, 3, 6). A general philosophy of knowledge was possible because although the sciences had distinct conceptual foundations all involved the dialectic of facts and ideas, the fundamental antithesis of philosophy.

METASCIENTIFIC CONFLICTS

Powell's conception of unity in the physical sciences was more ambitious than Whewell's. As mentioned earlier, Powell maintained a rigorous commitment to the principle of uniformity; but his understanding of unity was more than methodological – it entailed a vision of the union of all sciences under common, general laws. For Powell, the inductive principle pointed to 'the great archetype of *Unity*', since the tendency of scientific thought from the time of the ancient Greeks was away from isolated and disconnected branches towards 'harmony, simplicity, and unity of character'. This ideal was

[2] For another defence of natural history appealing to the unity or '*circle*', rather than the hierarchy of sciences, see Swainson 1834, 163–4. Whewell argued that the university should encourage natural history as a way of attracting the scientific interest of those not inclined to the 'severer forms of exact science'. See *Cambridge Philosophical Society* n.d., WP, R 18. 8[19].

the direction of history: 'All the first great modern advances were directed towards *combining* and *uniting* branches hitherto distinct, and tended to evince a *unity* of idea and principle' (Powell 1855, 41, 63).

Citing Faraday's work on the relationship between chemical and electrical action, he predicted a continuation of this unifying movement. Eventually, no essential distinction would be drawn between '*molecular* forces and those acting on matter in larger *masses*'; such a division was a temporary and provisional classification of disciplines. It was here that Powell made a direct criticism of Whewell, alleging that:

It is a reversal of the order of inductive advance to isolate each department of science, and to place it on a separate base, by a theory that would assign to each branch certain real differences of principle and peculiar fundamental ideas essentially characterising it. If such a distinction could be made out, it could be but a temporary and provisional ground of classification, in time to be superseded by a reduction to a higher common principle. (Powell 1855, 44–5; also 49)

On this view, there was no 'mysterious connection' of mechanics, as such, with the idea of causation. Rather, the success of this subject derived from its reduction of phenomena to 'simple and intelligible causes of force and motion'. As other disciplines began to emulate this pattern, Powell said that 'we bring those branches of science into the domain of exact science, and break down the line of demarcation which hitherto seemed to separate them' (Powell 1855, 45–6; also 1849, 35, for criticism of Whewell's distinction between statics and dynamics). This contradicted Whewell's claim that new truths were attained by the 'application of new ideas, not by the modification of old ones' (Whewell 1847a, II, 100–5; also 1854a, 30–1).

Powell and Whewell both used a general philosophy of science to make judgements about scientific research and its wider moral and theological implications. But their interpretations of the concept of the unity led to different recommendations. Most controversially, this was evident in their approach to the 'species question' and the nature of biology. Here Powell appealed to the notion of unity, asserting that the explanation of organic phenomena by natural causes lay within the legitimate scope of the life sciences. Whewell, however, argued that the history of science suggested that new domains of knowledge were associated with new fundamental ideas. Researchers had to resist the tendency to employ familiar causes and seek 'distinctly new conceptions' for 'newly studied phenomena' (Whewell 1847a, II, 100–5). Just as chemistry could not be reduced to mechanics, so physiological and biological phenomena could not be understood without reference to Final Cause – the peculiar idea appropriate to this area. In his natural theology, this acknowledgement of conceptual boundaries between the physical and organic sciences protected the values associated with the conviction that humanity held a special place in a designed universe. For Powell, on the other hand, such a position threatened the strategy of extending the compass of general laws of nature and claiming these as evidence of God's power. Here again, Whewell's philosophy was the enemy: it implied that disciplines such as physiology and biology 'involve altogether a *new class and order of ideas* of so peculiar a kind that they must stand out as entirely exceptional cases to the general *unity of the sciences*' (Powell 1855, 63).

The salient point here is that for Whewell, in contrast with Powell, the unity of science consisted in epistemological analogies, not substantive natural laws. There was a uniformity in the way knowledge was achieved, not a unity of the laws of nature. Although Whewell believed that the history of science displayed a 'consilience' of inductions towards simplicity and unity, he also stressed the particularity and integrity of different disciplines.

WHEWELL, POSITIVISM AND METASCIENCE

In 1865, J. S. Mill claimed that the essential elements of Comte's Positive philosophy were 'the general property of the age' (Mill 1866, 8). Robert Butts has proposed that Whewell, Herschel, and Mill contributed to this groundwork by arguing for the separation of science from religion and theology. In other words, they helped to bring about the situation desired and predicted by Comte: the intellectual autonomy of science.

Whewell, of course, is the unexpected inclusion here, since his philosophy of science was in some ways in competition with that of Herschel and Mill. Butts suggests, however, that like the others' Whewell's work called 'essentially for the epistemological insulation of science from other activities' (Butts 1985, 206). In one sense, this is a salutary corrective to the view, stemming partly from Huxley, that Whewell was the high priest of an anti-scientific and theologically supported movement.[3] As I have argued in earlier chapters, Whewell was often fighting against conservative religious forces in his effort to affirm the intellectual value of science. Butts makes the interesting point that by insisting that consilience of inductions delivered certain truths, Whewell left little room for the earlier doctrine of a 'holy alliance' between 'two truths' – the basis of natural theology. The result was that by the end of the century, very few scientists, unlike their earlier Victorian counterparts, felt any public pressure to reconcile their results with another source of belief.

It is fair to say that Whewell failed to stem this tide. However, this result did not occur because of a basic affinity between Whewell's programme and that of Mill's or Comte's, as Butts proposes. There were indeed some aspects of the philosophy of science that were the 'general property of the age' – recall, for example, Whewell's discussion with Herschel about the way in which they shared a vocabulary with Mill (ch. 4). But any further identification of Whewell's philosophy with the trend towards a Positivistic scenario for an autonomous science overlooks the way his project provided a resource for opponents of Positivism in Britain.

Whewell's insistence on the metaphysical foundations of scientific thought challenged both Comte's epistemology and his history of science. Since he stressed that scientific induction depended on metaphysical concepts, the success of science was not an argument for its epistemological insulation from other intellectual activities. While it is true that this message was weaker than the 'public science' of those who regarded any talk of metaphysics as a legacy of the earlier anti-scientific forces, it did find some listeners. In fact, it is here that Whewell gained a new audience, not among scientists, but among theologians and philosophers who sought to acknowledge the significance of science without accepting Positivism.

Whewell's philosophy of science was also an obstacle to those who wanted to use a Comtian interpretation of the unity of science as the justification for theological or social theories. Here there was a danger for Whewell, because Comte created the expectation that metascientific commentary would deliver a synthetic system embracing physical science and socio-political philosophy. This went beyond what Whewell was prepared to offer, and the finer inflections of his

---

[3] With Huxley in mind, Whewell remarked that it was sad that 'younger cultivators of science push out of sight and depreciate what their predecessors have done' (Whewell to Forbes, 24 July 1860, WP, O 15. 47[53]). See also Hodge 1991, 256–7.

philosophy of science – especially its implications for notions of unity –
meant that he lost other potential supporters.

SOME FAVOURABLE RESPONSES TO WHEWELL

In 1876 Henry Sidgwick and Mark Pattison reviewed the state of
philosophical studies in the two ancient English universities. They
were not pleased with what they found. But Sidgwick detected signs of
improvement and remarked that 'it is to Whewell more than to any
single man that the revival of Philosophy in Cambridge is to be
attributed' (Sidgwick 1876, 241–2; Pattison 1876). What did these
non-scientific writers admire in Whewell's work? No doubt it was
partly his campaign for the Moral Sciences Tripos, introduced in
1848, which recognized the value of subjects such as history, ethics,
and political economy in addition to mathematics and the classical
languages. But Sidgwick also referred to the way Whewell had
stimulated debate on the methodology of science; and Pattison, in an
earlier essay, welcomed Whewell's emphasis on the close relationship
between metaphysical thought and empirical research (in Pattison
1889, I, 440). These thinkers admired Whewell's work for the way it
kept science in contact with philosophy.[4]

From at least the 1850s, some leading theologians and philosophers
in Britain perceived an alliance between physical science and
positivist philosophy. In 1868 Archbishop William Thomson said
that this combination found strong support in England,

where the natural sciences are so eagerly pursued, where productive
industry is willing to push aside speculative inquiry in favour of the material
facts and observations which it needs for present use. (Thomson 1868, 6–7)

The Unitarian theologian, James Martineau, noticed something else.
In 1852 he suggested that there was an unexpected allegiance
between a conservative theology and Positivism, between

a Religion which exaggerates the functions and overstrains the validity of an
external authority, and a Science which deals only with objective facts,
perceived or imagined. (J. Martineau in 1879, 395)

The theology he referred to was the sophisticated critique of Henry
Mansel's Bampton lectures for 1858 on *The limits of religious thought*.
Drawing on William Hamilton's writings on the 'philosophy of the

---

[4] Sidgwick, however, criticized Whewell's moral philosophy. See R. Butts 1985, 208.

unconditioned', Mansel argued that it was impossible for the mind to reason philosophically about the notion of an absolute being separate from all the relations and conditions governing human thought. All efforts to conceive an infinite, absolute God proceeded 'from the consciousness, not of what is, but only of what is not'. But the inability of the mind to know the nature and attributes of the Deity did not imply non-existence. 'In this impotence of Reason', he suggested, 'we are compelled to take refuge in Faith and to believe that an Infinite Being exists, though we know not how' (Mansel 1859, 25, 75–80, 127).

Martineau detected an affinity between this theological 'nescience', as he called it, and the position of the English followers of Comte, such as Mill and Spencer. Both groups distrusted the metaphysical dimension of knowledge and the speculative power of the human mind. Thus a dangerous scenario was in place:

A Philosophy which surrenders all but phenomena is acceptable to those who would keep religion an affair of mere authority; and, for awhile, the strange but powerful partnership between negative metaphysics and theological dogma, will prevail against any attempt to reinstate the native trusts which have been so shaken in their seats. (J. Martineau 1866, iv)

One way to meet this situation was to affirm that abstract, speculative, metaphysical thought was essential to both science and theology. This was Martineau's reply. It was repeated by a variety of other writers, such as James McCosh, John Tulloch, Henry Calderwood, and John Daniel Morell, who did not necessarily agree on other religious or philosophical issues (see Yeo 1977). This answer was available to them in Whewell's *On the philosophy of discovery* of 1860.

This work was largely a republication of Book XII of the *Philosophy*, which was split into three separate works for the third edition. But it contained additional pieces, including two critical essays on Mansel and Comte, and three chapters on the theological bearing of the philosophy of science. These reasserted that metaphysical concepts were an integral part of both science and theology. Invoking his authority as an historian, Whewell insisted that progress of science was never marked, as Comte proposed, by the triumph of 'positive' thought over metaphysics. All major scientific discoveries were accompanied by intense discussion of the abstract concepts disallowed by Comte – for example, cause, force, atom, medium, final cause. Metaphysical discourse prepared the way for discoveries, as Comte admitted, but it also remained as an essential element: there was no science in which 'the most active disquisitions concerning ideas did not come *after*, not *before*, the first discovery of laws of phenomena' (Whewell 1860, 227). Mansel's arguments about the impotence of reason in dealing with abstract notions were also contradicted by the fact that mathematical science employed 'the notion of infinites, and leads to a great body of propositions concerning Infinites'. Moreover, when Newton spoke of God as an *'eternal'* Being, this language was 'not empty and unmeaning' (1860, 322, 325).

The other crucial intervention Whewell made was to detach Bacon from Comte. Positivist writers claimed Bacon's method as part of their heritage and celebrated Comte as successor to the position of Lord Chancellor of science. Whewell argued that Bacon was not a proponent of empiricist epistemology, that his campaign against the Scholastics for more empirical observation did not imply a denial of other elements of knowledge. Bacon recognized the importance of ideas; his concern with the confusion of terms in natural philosophy indicated an awareness of the role of clear metaphysics in the progress of science. Rather than being the antithesis to Plato, Bacon 'had the merit of showing that Facts and Ideas must be combined' (Whewell 1860, 145). Indeed, in this reinterpretation, Bacon personified the synthesis between the contesting forces in Whewell's history of epistemology: 'He held the balance, with no partial or feeble hand, between phenomena and ideas' (1860, 135). In a long article on Baconian scholarship for the *Edinburgh Review* in 1857, Whewell was able to cite the French writer, Charles Remusat, who now admitted that Bacon had been wrongly associated with the Positivist philosophy. Rightly understood, Bacon was an ally of those who recognized, in the words of Remusat, the 'worthlessness and emptiness of a merely empirical, mechanical, and utilitarian system of science' (Whewell 1857b, 321; Yeo 1985, 272-7).

Some writers believed Whewell's philosophy of science provided a basis for a renewed natural theology. Henry Manning may have had this in mind in 1845 when he declared:

Surely divine truth is susceptible, within the limits of revelation, of an expression and a proof as exact as the inductive sciences. Theology must be capable of a 'history and philosophy' if we had a Master of Trinity to write them. (Cited in Newsome 1966, 302)

Later, Adam Farrar proposed that the role of the intellectual faculties in religion could be understood by using the 'analysis which Dr. Whewell has given of their action in reference to science' (A. S. Farrar 1862, 39). However, there were dangers here. For example, John Daniel Morell told Whewell that he intended to apply 'the subjective element in Induction – as developed in your admirable treatise – to the method of theological research', in his forthcoming *Philosophy of religion* (Morell to Whewell, 20 December 1848, WP, Add. MS. c. 89[172]). Earlier, in his history of nineteenth-century philosophy, Morell claimed that the only hope for natural theology was a move from Paley's version to one founded on idealist epistemology:

Writers, for example, like McCulloch and Whewell, who have applied the highest scientific knowledge to maintain the validity of our natural religious conceptions, are philosophically speaking, most evidently idealistic in their tendency. (Morell 1847, II, 604)

But for Morell this meant an intuitional religion that made Revelation almost superfluous. Knowledge of God was a direct intuition, prior to any logical argument or reading of Scripture. Sedgwick spelt out the consequences here when he linked the German Idealist account of science with D. F. Strauss's denial of the historical foundation of Christianity: in seeking to derive all knowledge from a priori ideas these writers ignored the necessity for empirical observation in science and the importance of Revelation in religion (Morell 1849, 34–7, 85, 129–43, 210–15; Sedgwick 1850, cclxxi–cclxxiv; for responses to Morell, see Candlish 1849; Chalmers 1847).

Whewell was aware of the dangers of an inflated natural theology. Indeed, as John Brooke has indicated, he imposed limits on the design argument as early as his Bridgewater treatise, and affirmed the need for Revelation as a basis for moral knowledge (Brooke 1991a, 159–64). But in the *Philosophy of discovery* he did explore what he called the 'Theological Result of the Philosophy of Discovery' (Whewell 1860, 374). In short, this was the insight that the most profound evidence for natural theology derived not simply from the design or laws of nature, but from the success of science.

In 1860, as in his earlier *Of the plurality of worlds*, Whewell claimed that the manner in which the human mind gained knowledge of necessary truths about the world indicated an affinity between human and Divine minds. He was prepared to accept, with qualification, the Platonic view that the laws of the universe discovered by science were 'Ideas in the Mind of God' (Whewell 1860, 359).[5] Furthermore, the history of science revealed that this knowledge was progressive. 'What', asked Whewell, 'is the philosophical lesson to be derived from this progress, and from the new provinces thus added to human knowledge?' The answer was twofold. First, since science consisted in 'the idealization of facts', the progressive character of the knowledge suggested that humanity was acquiring a more comprehensive grasp of the Divine Ideas that governed Creation. Second, as new fundamental ideas emerged as the basis of new knowledge, 'we obtain a fresh proof of the Divine nature of the human mind'. But Whewell was quick to add that this was not the same as knowledge of God, of whom we 'can know *little*' (1860, 353, 374–6). Nevertheless, he was able to integrate the history of science into the Christian vision of redemption: human life on earth was not only a story of moral probation, but one of intellectual trial – the struggle to realize a potential for sharing in Divine truth (1860, 385–7, 398 and 1854c, 201–2; also Yeo 1979, 505–11 and 1986b, 279–81).

Whewell's work is usually seen as out of place in early Victorian culture because of its idealism and apparent affinity with Kant. But there is more to the reaction – of both admirers and critics – than this. As we have just seen, Whewell's idealism appealed to those who wanted to reject Positivism without abandoning science. It stressed the role of subjective and imaginative elements in scientific thought and hence offered an argument against sensationalist epistemology. James Martineau, for example, believed that Powell did not do this: his approach regarded the 'moral and spiritual world as a region not only distinct from, but disconnected with, the physical', and so effectively vitiated any 'link of transition from matter to mind' ([J. Martineau] 1855, 220). That is, in failing to assert the metaphysical foundation of science, Powell had surrendered the account of physical truth to the Positivists. John Tulloch agreed, saying that Powell had tried to tack Theism on to a positivist philosophy by 'a special reserve force of faith' (Tulloch 1868, 339). For these writers, Whewell not only discredited Positivism as a philosophy of science; he protected categories such as 'Mind' and 'Person' – in which there was theological investment – affirming that they were compatible with scientific knowledge. Into the bargain, he maintained that the success

[5] Compare Butts 1965 and Fisch 1985, 311–14 on the implications of this for Whewell's epistemology.

of science exploded the conservative nescience of Mansel (see Lightman 1987, 7–9, for Huxley's use of Mansel).

There were, however, other problems. Although some writers were happy with the way Whewell linked moral and physical knowledge, others disliked his position on the unity of science. Powell, as we have seen, was one of these. The issues in their dispute indicate that Whewell's idealism, *per se*, was by no means the major point of controversy. Rather, it was the way in which he disrupted the drive towards unification and synthesis, crucial to the projects of Powell, Comte, Chambers, Carpenter, and Spencer, and attractive to others, such as Babbage. In stressing the limits to unity dictated by the integrity of disciplines, Whewell ensured that his work did not simply confirm what Mill saw as the 'general property of the age'.

When writing about Comte, Whewell acknowledged that unity of explanation was a feasible goal. 'It may be', he said, 'that as science advances, all our knowledge may converge to one general and single aspect of the universe.' But he warned that this goal would not be reached if 'we refuse to admit those ideas which must be our stepping-stones' (Whewell 1860, 232–3). Indeed, Comte's classification of the sciences failed in this respect: it was necessarily artificial because it attempted to place physics and chemistry on a continuum that ignored the different fundamental ideas they involved (1860, 236). Although Whewell believed that his own inductive tables did reveal a 'constant advance towards unity, consistency, and simplicity', this applied only *within* different disciplines, such as Newtonian astronomy and optics (Whewell 1847a, II, 78). In fact, at present these were the only two subjects in which an extensive consilience of inductions had been reached, and this raised another caveat.

In his response to Mill's *System of logic* – published as *Of induction* in 1849 – Whewell advised against methodological policies hastily drawn from limited examples. He was referring to Mill's contention that the Baconian revolution was now complete and that future progress of science would stem from deduction, not induction. Whewell detected in this claim the erroneous Comtian notion that causal explanations from one area could be successfully extended to others, by deduction from established laws. He therefore urged that inspection of the present state of physical science disclosed 'a vast mass of cases' in which causes were not fully known. Knowledge of these 'new causes', and the proper generalization of the laws already known, could 'only be obtained by new *inductive* discoveries'. When Mill's book appeared, Whewell was also amazed that it relied so heavily on examples from 'Liebig's researches on physiological chemistry – *just published*'. He told Jones that not even the specialists could tell how much of this would 'stand as real discoveries' (Whewell to Jones, 7 April 1843, in Todhunter 1876, II, 313). After playing his strong card – a more detailed grasp of specific questions on the scientific agenda – Whewell charged that, in this instance, Mill's appeal to the unity of science had actually narrowed the field of 'scientific exertion' (Whewell 1860, 278–83, from a reprint of the 1849 text included in the 1860 book; also 1847a, II, 103).

In taking this stand against premature notions of unity, Whewell was regarded by some as lacking the synthetic scope of Comte and Spencer. There is a trace of this judgement in George Sarton's remark that Whewell's work was 'not a history of science as we know it today, but a juxtaposition of various special histories, which is something very different; it represents a lower stage of integration' (Sarton 1936, 63). Yet, as argued above, a recognition of the distinct epistemological foundations of different branches of science was central to Whewell's history and philosophy of science. It is ironic that Spencer was able to exploit this feature of Whewell's work against Comte in order to promote his own rival synthetic doctrine. In rejecting Comte's notion of 'a serial arrangement' of the sciences, Spencer cited Whewell's *History* in support of a more complex picture, without noting that it derived from this focus on the integrity of particular areas of inquiry (Spencer 1854, 152–3, 160).

This takes us back to Whewell's earlier diagnosis. In 1834 he had spoken of science as a 'great empire falling to pieces' under the impact of specialization and the accompanying institutional differentiation. But his mature works stressed that a unity of science could not be restored by ignoring the conceptual distinctiveness of its components. Furthermore, in spite of his reputation for omniscience, the moral of Whewell's writing was that the modern 'empire of Science' could not be held together by great individuals competent in several fields (Whewell 1847a, II, 4). Indeed, it was partly this situation that legitimated the metascientific role he assumed. While men of science – his contemporaries, such as Herschel, Sedgwick,

Faraday – were making discoveries in their chosen subjects, Whewell set himself the task of ensuring a philosophical overview of these activities.

The resulting picture was less ambitious than the vision of Comte and Spencer, or even Powell, but it was not without its power. In the commonwealth of sciences displayed by Whewell, the various branches of science had intellectual sovereignty over their respective domains:

The Mechanical, the Secondary Mechanical, the Chemical, the Classificatory, the Biological Sciences form so many Provinces in the Kingdom of knowledge, each in a great measure possessing its own peculiar fundamental principles. (Whewell 1847a, II, 19)

The role of the philosophy of science, and Whewell's own vocation, was to discern the analogies and distinctions in this grand intellectual structure. As noted earlier, Whewell often suggested that it might be possible to delineate a unity of epistemological process, while recognizing the integrity of different areas of inquiry. When explaining the advantage of taking the physical sciences as a starting point for his *Philosophy*, he observed that:

We have, at least, a definite problem before us. We have to examine the structure and scheme, not of a shapeless mass of incoherent materials, of which we doubt whether it be a ruin or a wilderness, but of a fair and lofty palace, still erect and tenanted, where hundreds of different apartments belong to a common plan. (Whewell 1847a, I, 14–15)

At the end of the first volume he averred that there was a 'Great Architect' who knew how 'Facts and Ideas' were intended to fit together in the 'lofty temple of Truth' (Whewell 1847a, I, 708). Nevertheless, Whewell's work stressed that the construction of the sciences was a human activity, one susceptible to delay and failure. Yet his analysis of the history and intellectual form of the sciences gave strong indications that they did indeed evince the 'common plan' preordained by God.

A conclusion at this point has the appeal of leaving Whewell as a metascientist who preserved, with certain qualifications, the idea of the unity of science. But this would be an inadequate account of the tensions in his work and his acknowledgement of them. In this and other chapters I have noted that whatever the scope of Whewell's ambitions, they must be set in the increasingly specialized world of science that he observed. In seeking to understand the dynamics of this progress of science through specialized disciplines, Whewell attempted to embrace, or at least monitor, a large number of them. And because he wanted to show that physical and moral (or social) knowledge was acquired in similar ways, his task was magnified. The span required exceeded the polymathic limits that allowed Herschel, De Morgan, and some other men of science to include the classics and modern poetry within their personal accomplishments.

None of this is surprising unless we take the view that Whewell's legendary omniscience transcended specialist boundaries. In fact, his work acknowledged specialization as the driving force of progress in science. We can find this acceptance in the very heart of his philosophy, as revealed in the crucial paper of 1844 on the fundamental antithesis of philosophy. As he had done in his two major works, Whewell here stressed the union of appropriate ideas and relevant empirical data as the condition of scientific theories capable of delivering necessary truths. But he also offered a view of 'the universal type of the progress of science'. This had two aspects. First, as fundamental ideas in particular sciences colligated more and more data, knowledge was 'reduced to principles gradually more simple'. Second, the progress of science also consisted in the expansion of the number of areas in which this kind of certain knowledge was possible. New fundamental ideas emerged as the organizing elements of newly explored subject matter. 'Such may soon be the case', he suggested, 'with the principles which are to be the basis of the philosophy of chemistry', or with electricity, galvanism, and magnetism (Whewell 1844, 12 quoted from a printed version, Cambridge 1844). Thus the move to simplicity in each mature science, as fundamental ideas were clarified, was countered by a diversification of scientific inquiry.[6]

Both these features of scientific development implied specialization. Whewell claimed that necessary truths could not be perceived in any science by those with 'indistinct conceptions', just as 'children and rude savages' were not reliably able to grasp them in geometry. Intense cultivation of particular sciences was essential for the clarification of the appropriate ideas, and hence 'persons new to chemical and classificatory science may not possess these ideas distinctly' (Whewell 1844, 2, 12). Thus the extension of scientific

---

[6] Whewell castigated the German idealist philosophers for presumptuously believing that they could see to the end of the philosophical drama stemming from the 'Fundamental Antithesis'. They had not considered the emergence of new fundamental ideas. See Whewell 1848a, 620.

knowledge over a wider domain – the idealization of more empirical phenomena – required close attention, by expert cultivators, to the ideal conceptions relevant to these new areas of study. The consequence was that the perception of necessary truth in some subjects might be confined to specialists.

The irony here is that Whewell's account of the 'universal type' of science endangered the scope of metascience. In stressing the integrity of disciplines, it revealed a situation in which the metascientist could not fully appreciate the thought of the specialist cultivators and might therefore be restricted in commenting on various issues. By 1860, with the Darwinian debates in mind, Whewell suggested that the subject had become too specialized for general discussion, telling Forbes that Bishop Wilberforce, in public debate with Huxley, was 'not prudent to venture into a field where no eloquence can supersede the need of precise knowledge'. Later, he told another correspondent that 'a person who ventures into the controversies which are at present agitated ought to have a great deal of specific knowledge, which I do not possess' (Whewell to Forbes, 24 July 1860, WP, O 15. 47[83]; Whewell to D. Brown, 26 October 1863, in Todhunter 1876, II, 434; on this famous clash, see Garland 1980, 106; Brooke 1991b, 41, 49–50).

Although he always anticipated the charge of trespass, Whewell assumed the role of critical commentator on the whole circle of sciences. His mathematical training and knowledge of the mature physical sciences gave him a secure basis; his extraordinary capacity, reading, and intellectual contacts, allowed him to appreciate the advances in chemistry, geology, physiology, and comparative anatomy. But his comments in the 1860s indicate that he saw the growth of the biological sciences as a major new domain demanding expert cultivation. In itself this did not disqualify the general account of the sciences he had produced, since this accommodated the intense specialization of new areas. But collectively, the breadth and depth of the sciences now placed informed, detailed comment on all, or even most of them, beyond the aspiration of the polymathic individual. Furthermore, in his review of Baconianism in 1857 Whewell raised the question of whether a metascientist could legislate for the specialist sciences: 'there is a legislation of scientific discovery which is wiser than any legislator, even than Bacon; namely, men of science themselves, employed in making discoveries, according to their own intellectual impulse' (Whewell 1857a, 302–3). This was almost an abdication of the metascientific role on the important topic of methodology, in favour of the judgements of practising experts.

But if Whewell did prophesy the demise of his own vocation he left an example of how philosophy of science could be pursued as part of wider moral and cultural engagements. When Mill recounted his own career as an attempt to expose the political implications of false philosophies of knowledge, he remembered Whewell as the most sophisticated representative of the 'notion that truths external to the mind may be known by intuition or consciousness, independently of observation and experience' (Mill 1971, 134). We now know that this is a caricature of Whewell's complex position, which did not ignore the empirical element of science; but it is also a biased analysis of the debates in which they participated. Mill charged that idealist epistemology was a defence of 'false doctrines and bad institutions'. But surely this is what Whewell was saying about the empiricism Mill upheld. Whewell did not conceal his fear that an empiricist philosophy of science, if widely accepted, would be a powerful support of utilitarian moral and social doctrines. This is why – apparently unnoticed by Mill – he also resisted the Tractarian theologians and their sympathizers who allowed physical science to be identified with these radical and secular positions.

Whewell knew, as well as Mill, that epistemology carried an ideological investment and, in particular, that accounts of science were contested by different groups with conflicting political interests. The conception of his metascientific project as a vocation depended on an explicit statement of his own commitments – to the reform of moral philosophy, to the affirmation of human reason, to the maintenance of the Anglican Church and its link with the State, and, most prominently, to the advocacy of science as the best example of human progress towards certain knowledge. Whewell did not present philosophy of science as a totally neutral discourse and was open about the presence of values in his own attempt to use it as a basis for reform in other areas. For a time, after his death and into the twentieth century, this candidness was lost.

References

UNPUBLISHED SOURCES

British Library, London – Babbage Papers; Macvey Napier Papers
Library of the Royal Society of London – John F. Herschel Papers
National Library of Scotland – Additional manuscripts: D. Brewster;
J. Lockhart
Trinity College Library, Cambridge – William Whewell Papers

PRINTED SOURCES

Note. I have used Houghton (1966–88) for the identification of authors of anonymous articles. I have not attempted to give a full listing of Whewell's publications; those included, like the other items, are cited in the text.

Aarsleff, H. (1971) 'Locke's reputation in nineteenth-century England', *Monist* 55: 392–422

Abercrombie, J. (1837) *The culture and discipline of the mind*, 6th edn, Edinburgh

Abir-Am, P. and D. Outram (eds.) (1987) *Uneasy careers and intimate lives: women in science, 1789–1979*, London

Agassi, J. (1971) 'Sir John Herschel's philosophy of success', *Historical Studies in the Physical Sciences*, 1: 1–36

Airy, G. B. (1896) *Autobiography of Sir George Biddell Airy*, Cambridge

Albert, Prince Consort (1859) 'Address', *Report of the 29th. meeting of the British Association for the Advancement of Science* (1860), London, lxix–lxxix

Allen, P. (1985) 'S. T. Coleridge's *Church and State* and the idea of an intellectual establishment', *Journal of the History of Ideas* 46: 89–107

Alter, P. (1987) *The reluctant patron: science and the state in Britain 1850–1920*, Berg

Altick, R. (1957) *The English common reader: a social history of the mass reading public*, Chicago

Annan, N. (1955) 'The intellectual aristocracy', in J. H. Plumb (ed.), *Studies in social history: a tribute to G. M. Trevelyan*, London, pp. 241–87

Anon. (1813) 'Advertisement' and 'Preface', *Annals of Philosophy* 1: ii–iv, 1–4
(1820) 'Education of the poor in France', *Edinburgh Review* 33: 494–509

(1828) 'Scientific education of the upper classes', *Westminster Review* 9: 328–73

(1831) 'Cambridge and Oxford education', *Westminster Review* 15: 56–69

(1834) 'On the application of the terms poetry, science and philosophy', *Monthly Repository of Theology and General Literature* 8: 323–331

(1843) *A letter to the Rev. William Whewell*, London

(1860) 'Whewell's philosophy of discovery', *Literary Gazette* 91: 24 March, 366

(1866) 'Obituary. The Reverend William Whewell', *The Athenaeum*, March, 333–4

Babbage, C. (1830) *Reflections on the decline of science in England*, London

(1835) *On the economy of machinery and manufactures*, 4th edn, enlarged, London

(1838) *Ninth Bridgewater treatise: a fragment*, London

(1851) *The Exposition of 1851; views of the industry, the science, and the government, of England*, London

Baily, F. (1833) 'A short account of some MSS letters . . .', *Report of the third meeting of the British Association for the Advancement of Science* (1834), London, 462–6

(1835) *An account of the Reverend John Flamsteed, the first Astronomer Royal*, London

[Baily, F.] (1836) 'Flamsteed, Newton, and Halley', *Magazine of Popular Science* 1: 83–96

(1837) *Supplement to the account of the Reverend John Flamsteed, with an author index*, London

Barrow, J. H. (1835) 'Account of the Rev. John Flamsteed', *Quarterly Review* 55: 96–128

Becher, H. W. (1980) 'William Whewell and Cambridge mathematics', *Historical Studies in the Physical Sciences* 11: 1–48

(1986) 'Voluntary science in nineteenth-century Cambridge university to the 1850s', *British Journal for the History of Science* 19: 57–87

(1991) 'William Whewell's odyssey: from mathematics to moral philosophy', in Fisch and Schaffer (eds.) (1991), pp. 1–29

Belsey, A. (1974) 'Interpreting Whewell', *Studies in History and Philosophy of Science* 5: 49–58

Berman, M. (1978) *Social change and scientific organization: the Royal Institution 1799–1844*, London

Biot, J. B. (1821) *Life of Sir Isaac Newton*, trans. H. Brougham (1829), London

Birks, T. R. (1834) *Oration on the analogy of mathematical and moral certainty*, Cambridge

Blair, A. (1824) 'Thoughts on some errors of opinion in respect to the advancement and diffusion of knowledge', *Blackwood's Magazine* 6: 26–33

Blanché, R. (1967) 'William Whewell', in P. Edwards (ed.), *Encyclopaedia of Philosophy*, 8 vols., New York

Bowden, J. W. (1834) 'The British Association', *Oxford University Magazine* 1: 401–12

(1839) 'The British Association for the Advancement of Science', *British Critic* 25: 1–48

Bowring, W. (1827) 'Education of the people', *Westminster Review* 7: 269–317

Brett-Smith, H. F. B. (ed.) (1947) *Peacock's four ages of poetry; Shelley's defence of poetry*, Oxford

Brewster, D. (1828) 'Recent history of astronomy', *Quarterly Review* 38: 1–15

(1830) 'Decline of science in England', *Quarterly Review* 43: 305–42

(1831a) 'Herschel's *Treatise on sound*', *Quarterly Review* 44: 475–511

(1831b) *The life of Sir Isaac Newton*, London

(1833–4) 'The British Scientific Association', *Edinburgh Review* 60: 363–94

(1834) 'The Bridgewater bequest: Whewell's Astronomy and general physics', *Edinburgh Review* 58: 422–57

(1837) 'Whewell's History of the inductive sciences', *Edinburgh Review* 66: 110–51

(1838a) 'M. Comte's Course of positive philosophy', *Edinburgh Review* 67: 271–308

(1838b) 'Weather almanacks – the late frost', *Monthly Magazine* 1: 76–84

(1839) 'The sciences connected with natural theology', *Monthly Chronicle* 3: 97–115

(1842a) 'Whewell's Philosophy of the inductive sciences', *Edinburgh Review* 74: 265–306

(1842b) 'The Encyclopaedia Britannica', *Quarterly Review* 70: 44–72

(1845) 'Vestiges of the natural history of creation', *North British Review* 3: 470–515

(1855) *Memoirs of the life, writings, and discoveries of Sir Isaac Newton*, 2 vols., London

(1858) 'Researches on light', *North British Review* 29: 178–210

Brock, W. (1988) 'British science periodicals and culture: 1820–1850', *Victorian Periodicals Review* 21: 47–55

Brooke, J. H. (1977) 'Natural theology and the plurality of worlds: observations on the Brewster-Whewell debate', *Annals of Science* 34: 221–86

(1987) 'Joseph Priestley (1733–1804) and William Whewell (1794–1866): apologists and historians of science. A tale of two stereotypes', in R. Anderson and C. Lawrence (eds.), *Science, medicine and dissent: Joseph Priestley (1733–1804)*, London

(1991a) 'Indications of a creator: Whewell as apologist and priest', in Fisch and Schaffer (eds.) (1991), pp. 149–73

(1991b) *Science and religion: some historical perspectives*, Cambridge

Brougham, H. (1826) 'Diffusion of knowledge', *Edinburgh Review* 45: 189–99

(1827a) *Discourse on the objects, advantages, and pleasures of science*, London

(1827b) 'Royal society – president's discourses', *Edinburgh Review* 46: 352–65

Buchdahl, G. (1991) 'Deductivist versus inductivist approaches in the philosophy of science as illustrated by some controversies between Whewell and Mill', in Fisch and Schaffer (eds.) (1991), pp. 311–44

Bulwer-Lytton, E. (1833) *England and the English*, 2 vols., London

Burckhardt, J. (1954) [1860] *The civilization of the Renaissance in Italy*, introd. H. Holborn, New York

Burke, P. and R. Porter (eds.) (1987) *The social history of language*, Cambridge

Burrow, J. W. (1981) *A liberal descent. Victorian historians and the English past*, Cambridge

Bury, J. P. T. (ed.) (1967) *Romilly's Cambridge diary, 1832–42: selected passages from the diary of the Rev. Joseph Romilly, fellow of Trinity College and Registrar of the University of Cambridge*, Cambridge

[Butler, A.] (1841) 'Whewell's philosophy of the inductive sciences', *Dublin University Magazine* 17: 194–211; 555–72

Butterfield, H. (1931) *The Whig interpretation of history*, London

Butts, R. E. (1965) 'Necessary truth in Whewell's theory of science', *American Philosophical Quarterly* 2: 161–81

  (1968) *William Whewell's theory of scientific method*, Pittsburgh

  (1985) '"A purely scientific temper": Victorian expressions of the ideal of an autonomous science', in N. Rescher (ed.), *Reason and rationality in natural science*, New York, pp. 191–213

Candlish, T. (1849) 'Morell's philosophy of religion', *North British Review* 11: 1–43; 293–336

Cannon, S. F. [W. F.] (1961) 'John Herschel and the idea of science', *Journal of the History of Ideas* 22: 215–39

  [W. F.] (1964) 'William Whewell: contributions to science and learning', *Notes and Records of the Royal Society* 19: 176–91

  (1978) *Science in culture: the early Victorian period*, New York

Cantor, G. N. (1982) 'The eighteenth-century problem', *History of Science* 20: 44–63

  (1983) *Optics after Newton: theories of light in Britain and Ireland, 1704–1840*, Manchester, pp. 1–3

  (1991a) 'Between rationalism and romanticism: Whewell's historiography of the inductive sciences', in Fisch and Schaffer (eds.) (1991), pp. 67–86

  (1991b) *Michael Faraday: Sandemanian and scientist: a study of science and religion in the nineteenth century*, London

Carlisle, H. (1882) 'William Whewell', *MacMillan's Magazine* 45: 138–44

Carlyle, T. (1829) 'Signs of the times', *Edinburgh Review* 49: 439–59

Chalmers, A. (1991) *Science and its fabrications*, Milton Keynes

Chalmers, T. (1836–42) *The works of Thomas Chalmers*, 25 vols. Glasgow

  (1847) 'Morell's modern philosophy', *North British Review* 6: 271–331

Chambers, R. (1844) *Vestiges of the natural history of creation*, London

  (1846) *Explanations: a sequel to vestiges of the natural history of creation*, London

Checkland, S. G. (1949) 'The propagation of Ricardian economics in

England', *Economist* 16: 40–56

(1951) 'The advent of academic economics in England', *The Manchester School of Economic and Social Studies* 19: 43–70

Chenevix, R. (1820) 'State of science in England and France', *Edinburgh Review* 34: 383–422

Clark, J. W. and T. M. Hughes (1890) *The life and letters of the Rev. Adam Sedgwick*, 2 vols., Cambridge

Cockburn, H. (1874) *Life and correspondence of Francis Jeffrey*, Edinburgh

Cockburn, W. (1838) *A remonstrance, addressed to his grace the Duke of Northumberland, upon the dangers of peripatetic philosophy*, London

(1845) *The Bible defended against the British Association*, 5th edn, London

Cohen, I. B. (1985) *Revolution in science*, Cambridge, Mass

Coleridge, S. T. (1817) 'General introduction; or, preliminary treatise on method', *Encyclopaedia Metropolitana* 1: 1–43, London

(1972) [1830] *On the constitution of church state according to the idea of each*, ed. J. Barrell

Collini, S., D. Winch, and J. Burrow (1983) *That noble science of politics. A study in nineteenth-century intellectual history*, Cambridge

Comte, A. (1970) [1830] *Introduction to positive philosophy*, trans. F. Ferre, Indianapolis

Cooter, R. (1984) *The cultural meaning of popular science: phrenology and the organization of consent in nineteenth-century Britain*, Cambridge

Copleston, E. (1807) *Advice to a young reviewer, with a specimen of the art*, Oxford

(1810) *A reply to the calumnies of the Edinburgh Review against Oxford, containing an account of the studies pursued in that university*, Oxford

Corrigan, T. (1980) '*Biographia literaria* and the language of science', *Journal of the History of Ideas* 41: 399–419

Corsi, P. (1987) 'The heritage of Dugald Stewart: Oxford philosophy and the method of political economy, 1809–1832', *Nuncius* 2: 89–144

(1988) *Science and religion: Baden Powell and the Anglican debate 1800–1860*, Cambridge

Crosland, M. (ed.) (1975) *The emergence of science in western Europe*, London

Cruse, A. (1930) *The Englishman and his books in the early nineteenth century*, London

Cunningham, A. (1988) 'Getting the game right: some plain words on the identity and invention of science', *Studies in History and Philosophy of Science*, 19: 365–89

Cunningham, A. and N. Jardine (eds.) (1990) *Romanticism and the sciences*, Cambridge

Davie, G. (1964) *The democratic intellect: Scotland and her universities in the nineteenth century*, Edinburgh

De Morgan, A. (1832a) 'Study of natural philosophy', *Quarterly Journal of Education* 3: 60–73

(1832b) 'State of mathematical and physical sciences at Oxford', *Quarterly Journal of Education* 4: 191–208

(1835) 'English science', *British and Foreign Review* 1: 134–57
(1837) 'Theory of probabilities – part II', *Dublin Review* 3: 237–48
(1840) 'Philosophy of the inductive sciences', *The Athenaeum* no. 672, 12 September, 707–9
(1842) 'Science and rank', *Dublin Review* 13: 413–48
(1845) 'Speculators and speculations', *Dublin Review* 19: 99–129
(1855) 'Sir David Brewster's *Life of Newton*', *North British Review* 23: 307–38
(1914) [1846] 'Newton', in *Essays on the life and work of Isaac Newton*, ed. P. Jourdain, Chicago, pp. 3–63
(1915) [1871] *A budget of paradoxes*, 2nd edn, 2 vols., New York
DeQuincey, T. (1824) 'Superficial knowledge', *London Magazine* 10: 25–8
Desmond, A. (1989) *The politics of evolution: morphology, medicine, and reform in radical London*, Chicago
Dickinson, H. W. (1932) 'J. O. Halliwell and the historical society of science', *Isis* 18: 127–32
Don Vann, J. and R. Van Arsdel (eds.) (1989) *Victorian periodicals: a guide to research*, 2 vols., New York
Drinkwater, J. (1833) 'The life of Kepler', in *Lives of eminent persons*, London
Eagleton, T. (1984) *The function of criticism from 'the Spectator' to post-structuralism*, London
Ebison, M. (ed.) (1971) *The harvest of a quiet eye. A selection of scientific quotations by Alan L. Mackay*, Bristol and London
Edgeworth, M. (1971) *Letters from England 1831–44*, ed. C. Colvin, London
Elkana, Y. (ed.) (1984) *William Whewell: selected writings on the history of science*, Chicago
Emerson, R. L. (1988) 'The Scottish Enlightenment and the end of the Philosophical Society of Edinburgh', *British Journal for the History of Science* 21: 33–66
Ezrahi, Y. (1990) *The descent of Icarus: science and the transformation of contemporary democracy*, Cambridge, Mass.
Farrar, A. S. (1862) *A critical history of free thought in reference to the Christian religion*, London
Farrar, W. V. (1975) 'Science and the German university system, 1790–1850', in Crosland (ed.), (1975) pp. 179–92
Fisch, M. (1985) 'Necessary and contingent truth in William Whewell's antithetical theory of knowledge', *Studies in History and Philosophy of Science*, 16: 275–314
(1991a) *William Whewell. Philosopher of science*, Oxford
(1991b) 'A philosopher's coming of age: a study in erotetic intellectual history', in Fisch and Schaffer (eds.) (1991), pp. 31–66
Fisch, M. and S. Schaffer (eds.) (1991) *William Whewell. A composite portrait*, Oxford
Forbes, D. (1952) *The liberal Anglican idea of history*, Cambridge
Forbes, J. D. (1834) 'Address', *Report of the fourth meeting of the British*

Association for the Advancement of Science (1835), London, xi–xxiii

(1849) *The danger of superficial knowledge*, London

(1858) 'The history of science; and some of its lessons', *Fraser's Magazine* 57: 283–94

Furst, L. (ed.) (1980) *European romanticism. Self definition*, New York

Galloway, T. (1833–4) 'Sir John Herschel's *Astronomy*', *Edinburgh Review* 58: 164–98

(1836) 'Life and observations of Flamsteed', *Edinburgh Review* 62: 359–97

(1844) 'The martyrs of science', *Edinburgh Review* 80: 164–98

Galton, F. (1874) *English men of science: their nature and nurture*, London

(1892) [1869] *Hereditary genius: an inquiry into its laws and consequences*, 2nd edn, London

Garland, M. M. (1980) *Cambridge before Darwin: the ideal of a liberal education 1800–1860*, Cambridge

Gascoigne, J. (1984) 'Mathematics and meritocracy: the emergence of the Cambridge mathematical tripos', *Social Studies of Science* 14: 547–84

(1989) *Cambridge in the age of the Enlightenment: science, religion and politics from the Restoration to the French Revolution*, Cambridge

(1990) 'A reappraisal of the role of the universities in the scientific revolution', in D. C. Lindberg and R. S. Westman (eds.), *Reappraisals of the scientific revolution*, Cambridge, pp. 207–60

Gascoigne, R. M. (1984) *A historical catalogue of scientists and scientific books from the earliest times to the close of the nineteenth century*, New York

Glisserman, S. (1975) 'Early Victorian science writers and Tennyson's "In Memoriam": a study in cultural exchange', *Victorian Studies* 19: 277–308

Graham, L., W. Lepenies and P. Weingart (eds.) (1983) *Functions and uses of disciplinary histories*, Dordrecht

Graves, R. P. (1882–5) *Life of Sir William Rowan Hamilton*, 3 vols. Dublin

Griggs, E. L. (ed.) (1956–71) *Collected letters of Samuel Taylor Coleridge*, 6 vols., Oxford

Gross, J. (1969) *The rise and fall of the man of letters: aspects of English literary life since 1800*, London

Grove, W. (1843) 'Physical science in England', *Blackwood's Magazine* 54: 514–25

Habermas, J. (1970) *Towards a rational society*, London

(1974) 'The public sphere: an encyclopedia article', *New German Critique* 3: 49–55

(1989) [1962] *The structural transformation of the public sphere: an inquiry into a category of bourgeois society*, trans. T. Burger with the assistance of F. Lawrence, Cambridge, Mass.

Hahn, R. (1971) *Anatomy of a scientific institution: the Paris Academy of Sciences, 1866–1803*, Berkeley

Hall, M. B. (1984) *All scientists now: the Royal Society in the nineteenth century*, Cambridge

Hamilton, W. (1831a) 'Universities of England: Oxford', *Edinburgh Review*

53: 384–427

(1831b) 'English universities: Oxford', *Edinburgh Review* 54: 478–504

(1836) *Letter to the right honorable The Lord Provost, magistrates, and town council, patrons of the University of Edinburgh*, Edinburgh

Hankins, T. L. (1980) *Sir William Rowan Hamilton*, Baltimore and London

Harcourt, W. V. (1831) 'Address', *Report of the first and second meetings of the British Association for the Advancement of Science* (1833), London, 17–41

Herschel, J. (1830) *A preliminary discourse on the study of natural philosophy*, London

Herschel, J.(*c.* 1830) 'Sound', *Encyclopaedia Metropolitana* 4: 747–824

(1841) 'Whewell on inductive sciences', *Quarterly Review* 68: 177–238

(1845) 'Address', *Report of the fifteenth meeting of the British Association for the Advancement of Science* (1846), London, xxvii–xliv

(1848) 'Humboldt's *Kosmos*', *Edinburgh Review* 87: 170–228

(1849) *The Admiralty manual of scientific enquiry*, London

(1857) *Essays from the Edinburgh and Quarterly Reviews*, London

(1867–8) 'The Reverend William Whewell DD', *Proceedings of the Royal Society of London* 135: li–lxi

Heyck, T. (1982) *The transformation of intellectual life in Victorian England*, London

Hill, A. G. (ed.) (1978–9) *The letters of William and Dorothy Wordsworth*, 2nd edn, from the first edition edited by the late E. de Selincourt, vols. 4 and 5, Oxford

Hilton, B. (1988) *The age of atonement: the influence of evangelicalism on social and economic thought 1795–1865*, Oxford

Hodge, M. J. S. (1991) 'The history of the earth, life, and man: Whewell and paletiological science', in Fisch and Schaffer (eds.) (1991), pp. 255–88

Hohendahl, P. U. (1982) *The institution of criticism*, Ithaca

Hornberger, T. (1949) 'Halliwell-Phillips and the history of science', *Huntingdon Library Quarterly* 12: 391–99

Houghton, W. (1966–88) *The Wellesley index to Victorian periodicals*, 5 vols., Toronto

(1982) 'Periodical literature and the articulate classes', in Shattock and Wolff (eds.) (1982), pp. 3–27

Hume, D. (1948) [1779] *Dialogues concerning natural religion*, ed. H. D. Aiken, New York

Husserl, E. (1970) *The crisis of European sciences and transcendental phenomenology: an introduction to phenomenological philosophy*, trans. with introduction by D. Carr, Evanston

Huxley, T. H. (1894) 'Past and present', *Nature* 51: 1–3

Inkster, I. (1979) 'London science and the seditious meetings act of 1817', *British Journal for the History of Science* 12: 192–6

Jeffrey, F. (1810) 'Stewart's philosophical essays', *Edinburgh Review* 17: 167–211

Jones, R. (1831) *An essay on the distribution of wealth and on the sources of taxation*,

London
(1859) *Literary remains*, ed. with an introduction by W. Whewell, London
Jones, R. F. (1961) *Ancients and moderns: a study of the rise of the scientific movement in England*, 2nd edn, Gloucester, Mass.
Kant, I. (1933) *Critique of pure reason*, trans. N. Kemp Smith, London
(1949) *Critique of practical reason and other writings in moral philosophy*, ed. L. Beck, Chicago
Kelly, T. (1966) *Early public libraries: a history of public libraries in Great Britain before 1850*, London
Kent, C. (1978) *Brains and numbers: elitism, Comtism, and democracy in mid-Victorian England*, Toronto
(1989) 'Victorian periodicals and the constructing of reality', in Don Vann and Van Arsdel (eds.) (1989), pp. 1–12
Knight, D. M. (1967) 'The scientist as sage', *Studies in Romanticism* 6: 65–88
(1986) *The age of science: the scientific world-view in the nineteenth century*, New York
Kuhn, T. S. (1977) *The essential tension: selected studies in scientific tradition*, Chicago
Lang, A. (1897) *Life and letters of John Gibson Lockhart*, 2 vols., London
Laudan, L. (1968) 'Theories of scientific method from Plato to Mach: a bibliographical review', *History of Science* 7: 1–63
(1971) 'William Whewell on the consilience of inductions', *Monist* 55: 368–9
(1981) *Science and hypothesis: historical essays on scientific methodology*, Dordrecht
Lepenies, W. (1988) *Between literature and science: the rise of sociology*, Cambridge
Levere, T. H. (1981) *Poetry realized in nature: Samuel Taylor Coleridge and early-nineteenth-century science*, New York
(1989) '"The lovely shapes and sounds intelligible": Samuel Taylor Coleridge, Humphry Davy, science and poetry', in J. Christie and S. Shuttleworth (eds.), *Nature transfigured: science and literature, 1700–1900*, Manchester
Levine, P. (1986) *The amateur and the professional: antiquarians, historians, and archaeologists in Victorian England, 1838–1886*, Cambridge
Lightfoot, J. B. (1866) *In memory of William Whewell DD. A sermon preached in the college chapel, March 18th, 1866*, London
Lightman, B. (1987) *The origins of agnosticism: Victorian unbelief and the limits of knowledge*, Baltimore and London
Long, G. (1841) *An essay on the moral nature of man*, London
Losee, J. (1983) 'Whewell and Mill on the relation between philosophy of science and history of science', *Studies in History and Philosophy of Science* 14: 113–26
Lyell, C. (1826) 'Scientific institutions', *Quarterly Review* 34: 153–79
(1827) 'State of the universities', *Quarterly Review* 36: 216–68

(1845) *Travels in North America*, 2 vols., London

Lyell, K. M. (ed.) (1881) *Life, letters and journals of Sir Charles Lyell*, 2 vols., London

Macaulay, T. B. (1913–15) *History of England from the accession of James II*, ed. C. Firth, 6 vols., London

McEvoy, J. G. and J. E. McGuire (1975) 'God and nature: Priestley's way of rational dissent', *Historical Studies in the Physical Sciences* 6: 325–404

Mackintosh, J. (1836) *Dissertation on the progress of ethical philosophy*, ed. with a preface by W. Whewell, Edinburgh

MacLeod, R. M. (1972) 'The resources of science in England: the endowment of science movement, 1868–1900', in P. Mathias (ed.), *Science and society, 1600–1900*, Cambridge, pp. 111–166

(1983) 'Whigs and savants: reflections on the reform movement in the Royal Society, 1830–48', in I. Inkster and J. Morrell (eds.), *Metropolis and province: science in British culture, 1780–1850*, London, pp. 55–90

Mansbridge, A. (1923) *The older universities of England*, Oxford and Cambridge

Mansel, H. L. (1859) *The limits of religious thought examined*, 3rd edn, London

Martineau, H. (1875) *The positive philosophy of Auguste Comte*, 2nd edn, 2 vols., London

[Martineau, J.] (1855) 'Contemporary literature: theology and philosophy', *Westminster Review* 64: 205–25

(1866) *Essays, philosophical and theological*, London

(1879) *Studies of Christianity*, London

Maurer, O. (1948) 'Anonymity vs signature in Victorian reviewing', *Studies in English* 28: 1–27

Maxwell, J. C. (1987) 'Whewell's writings and correspondence', *Nature*, 6 July, 206–8

(1890) *Scientific papers*, ed. W. D. Niven, 2 vols. Cambridge

Mendelsohn, E. (1964) 'The emergence of science as a profession in nineteenth-century Europe', in K. Hill (ed.), *The management of scientists*, Boston, pp. 3–48

Merton, R. K. (1965) *On the shoulders of giants: a Shandean postscript*, New York

(1973) 'The normative structure of science', in R.K. Merton, *The sociology of science*, ed. N. W. Storer, Chicago and London, pp. 267–78

Merz, J. T. (1896–1904) *A History of European thought in the nineteenth century*, 4 vols., Edinburgh and London

Mill, J. S. (1831) 'Herschel's Discourse', *Examiner*, 20 March, 179–80

(1833) 'Thoughts on poetry and its varieties', in Robson and Stillinger (eds.), vol. i, pp. 341–65

(1835) 'Sedgwick's Discourse', in Robson and Stillinger (eds.) (1981–91), vol. x, pp. 33–74

(1836) 'On the definition of political economy, and the method of philosophical investigation in that science', *London and Westminster Review* 26: 1–29

(1856) [1843] *A system of logic ratiocinative and inductive: being a connected view of the principles of evidence and the methods of scientific investigation*, 4th edn, London

(1852) 'Doctor Whewell on moral philosophy', *Westminster Review* 58: 349–85

(1859) *Dissertations and discussions*, 2 vols., London

(1866) *Auguste Comte and positivism*, London

(1971) [1873] *Autobiography*, ed. J. Stillinger, Oxford

(1980) *Mill on Bentham and Coleridge*, ed. F. R. Leavis, Cambridge

Miller, D. P. (1983) 'Between hostile camps: Sir Humphry Davy's Presidency of the Royal Society of London 1820–1827', *British Journal for the History of Science* 16: 1–47

Moll, G. (1831) *On the alleged decline of science in England*, London

Monk, W. (ed.) (1972) *The journals of Caroline Fox 1835–1871: a selection*, London

Moore, J. R. (1985) 'Darwin of Down', in D. Kohn (ed.), *The Darwinian heritage*, Princeton

(ed.) (1989) *History, humanity and evolution: essays for John C. Greene*, Cambridge

Morell, J. D. (1847) *An historical and critical view of the speculative philosophy of Europe in the nineteenth century*, 2nd edn, 2 vols., London

Morley, E. J. (ed.) (1927) *The correspondence of H. C. Robinson and the Wordsworth circle (1808–66)*, 2 vols. Oxford

Morrell, J. B. (1971a) 'Professors Robison and Playfair, and the *Theophobia gallica*: natural philosophy, religion and politics in Edinburgh, 1789–1815', *Notes and Records of the Royal Society* 16: 43–63

(1971b) 'Individualism and the structure of British science in 1830', *Historical Studies in the Physical Sciences* 3: 183–204

(1984) 'Brewster and the early British Association for the Advancement of Science', in Morrison-Low and Christie (eds.) (1984), pp. 25–30

(1992) 'The judge and purifier of all', *History of Science* 30, 97–113

Morrell, J. B. and A. Thackray (1981) *Gentlemen of science: early years of the British Association for the Advancement of Science*, Oxford

(1984) *Gentlemen of science: early correspondence of the British Association for the Advancement of Science*, London

Morrison-Low, A. D. and J. J. R. Christie (eds.) (1984) *Martyr of science: Sir David Brewster 1781–1868*, Edinburgh

Munby, A. N. L. (1968) *The history and bibliography of science in England: the first phase, 1833–1845*, Berkeley

Napier, M. (ed.) (1879) *Selections from the correspondence of the late Macvey Napier*, London

Newman, F. (ed.) (1843) *The English universities, from the German of V. A. Huber*, 2 vols., London

Newman, J. H. (1841) *The Tamworth reading room: letters on an address delivered by Sir Robert Peel*, London

Newsome, D. (1966) *The parting of friends: a study of the Wilberforces and Henry Manning*, London

Olby, R., G. Cantor, J. Christie and M. Hodge (eds.) (1990) *Companion to the history of modern science*, London

Oldroyd, D. R. (1980) 'Sir Archibald Geike and the problem of Whig historiography', *Annals of Science* 37: 441–62

Outram, D. (1976) 'Scientific biography and the case of Georges Cuvier: with a critical bibliography', *History of Science* 14: 101–37

(1978) 'The language of natural power: the funeral *éloges* of Georges Cuvier', *History of Science* 16: 153–178

(1980) 'Politics and vocation: French science, 1793–1830,' *British Journal for the History of Science* 13: 27–43

(1984) *Georges Cuvier: vocation, science and authority in post-revolutionary France*, Manchester

(1986) 'Uncertain legislator: Georges Cuvier's laws of nature in their intellectual context', *Journal of the History of Biology* 19: 323–68

Paley, W. (1809) *The principles of moral and political philosophy*, 17th edn, 2 vols., London

Paradis, J. and T. Postlewait (eds.) (1981) *Victorian science and Victorian values*, *Annals of the New York Academy of Sciences*, vol. 360, New York

Passmore, J. (1966) *A hundred years of philosophy*, Harmondsworth

Patterson, E. (1969) 'Mary Somerville, FRS', *British Journal of the History of Science* 4: 311–29

Pattison, M. (1876) 'Philosophy at Oxford', *Mind* 1: 82–97

(1889) *Essays*, ed. H. Nettleship, 2 vols., Oxford

Paul, C. B. (1980) *Science and immortality: the éloges of the Paris Academy 1699–1791*, Berkeley

Peacock, T. (1893) *Headlong Hall*, ed. R. Garnett, London

Playfair, J. (1808) 'La Place, *Traité de Méchanique Céleste*', *Edinburgh Review* 11: 249–84

(1810), 'Laplace's *System of the World*', *Edinburgh Review* 10: 396–417

(1822) *The works of John Playfair*, 4 vols., Edinburgh

Porter, R. (1976) 'Charles Lyell and the principles of the history of geology', *British Journal for the History of Science* 9: 91–103

(1982) 'Charles Lyell: the private and public faces of science', *Janus* 69: 29–50

Powell, B. (1834a) *An historical view of the progress of the physical and mathematical sciences, from the earliest ages to the present times*, London

(1834b) 'Physical studies in Oxford', *Quarterly Journal of Education* 7: 47–54

(1837) *On the nature and evidence of the primary laws of motion*, Oxford

(1838) *The connexion of natural and divine truth*, London

(1839) 'M. Comte's system of positive philosophy', *Monthly Chronicle* 3: 227–38

(1841) *A general and elementary view of the undulatory theory*, London

(1849) *On necessary and contingent truth*, Oxford

(1850) 'Letter to Doctor Whewell', *Philosophical Magazine*, series 3, 36: 235
(1855) *Essays on the inductive philosophy, the unity of worlds and the philosophy of creation*, London
(1856) 'Sir Isaac Newton', *Edinburgh Review* 103: 499–535
(1859) *The order of nature considered in reference to the claims of revelation*, London
Preyer, R. (1981) 'The romantic tide reaches Trinity', in Paradis and Postlewait (eds.), pp. 39–68
Priestley, J. (1775) *The history and present state of electricity with original experiments*, 3rd edn, 2 vols., London
Pusey, P. (1838) 'Plato, Bacon and Bentham', *Quarterly Review* 61: 462–506
Pym, H. N. (1882) *Memories of old friends, being abstracts from the journals and letters of Caroline Fox of Penjerric, Cornwall from 1835 to 1871*, London
Rashid, S. (1977) 'Richard Whately and Christian political economy at Oxford', *Journal of the History of Ideas* 38: 147–55
Raverat, G. (1967) *Period piece: a Cambridge childhood*, London
Ravetz, J. (1984) 'Ideological commitments in the philosophy of science', *Radical Philosophy*, summer: 5–11
Rennell, T. (1819) *Remarks on scepticism*, 3rd edn, London
Richards, E. (1989) 'Huxley and woman's place in science: the "woman question" and the control of Victorian anthropology', in Moore (ed.), pp. 253–84
Richards, J. (1988) *Mathematical visions: the pursuit of geometry in Victorian England*, Boston
Rigaud, S. P. (1836) 'Newton and Flamsteed', *Edinburgh Philosophical Magazine* 8: 138–47
(1838) *Historical essay on the first publication of Sir Isaac Newton's Principia*, Oxford
(1851) *Correspondence of scientific men of the seventeenth century*, 2 vols., Oxford
Rimmer, W. G. (1960) *Marshalls of Leeds, flax spinners 1788–1886*, Cambridge
Roach, J. P. C. (1959) 'Victorian universities and the national intelligentsia', *Victorian Studies* 3: 131–150
Robinson, H. C. (1872) *Diary, reminiscences, and correspondence of Henry Crabb Robinson*, ed. T. Sadler, 3rd edn, 2 vols., London
Robison, J. (1797) 'Physics', *Encyclopaedia Britannica* XIV: 637–59, 3rd edn., Edinburgh
Robson, J. and J. Stillinger (eds.) (1981–91) *Collected works of John Stuart Mill*, 33 vols., Toronto
Robson, R. (1964) 'William Whewell, FRS: academic life', *Notes and Records the Royal Society of London*, 19: 168–76
(1967) 'Trinity College in the age of Peel', in R. Robson (ed.), *Ideas and institutions of Victorian Britain*, London, pp. 168–76
Rose, H. J. (1826) *The tendency of prevalent opinions about knowledge considered*, Cambridge
(1834) *An apology for the study of divinity*, London

Rosen, E. (1964) 'Renaissance science as seen by Burckhardt and his successors', in T. Helton (ed.), *The Renaissance: a reconsideration of the theories and interpretations of the age*, pp. 77–104

Ross, S. (1961) 'Faraday consults the scholars: the origins of the terms of electrochemistry', *Notes and Records of the Royal Society* 16: 187–220

(1962) '"Scientist": the story of a word', *Annals of Science* 18: 65–85

(1978) 'John Herschel on Faraday and on science', *Notes and Records of the Royal Society of London* 33: 77–82

Rothblatt, S. (1976) *Tradition and change in English liberal education: an essay in history and culture*, London

(1981) [1968] *The Revolution of the dons: Cambridge and society in Victorian England*, 2nd edn, Cambridge

(1985) 'The notions of an open scientific community in historical perspective', in M. Gibbons and B. Wittrock (eds.), *Science as a commodity: threats to the open community of scholars*, London, pp. 21–75

Rudwick, M. J. S. (1985) *The great Devonian controversy: the shaping of scientific knowledge among gentlemanly specialists*, Chicago

Rupke, N. (1983) *The great chain of history: William Buckland and the English school of geology*, Oxford

Ruse, M. (1976) 'The scientific methodology of William Whewell', *Centaurus* 20: 127–57

(1977) 'William Whewell and the argument from design', *Monist* 60: 244–68

(1991) 'William Whewell: omniscientist', in Fisch and Schaffer (eds.), pp. 87–116

Sanderson, M. (1972) *The universities and British industry 1850–1970*, London

Sarton, G. (1936) *The study of the history of science*, Cambridge, Mass.

Schaffer, S. (1986) 'Scientific discoveries and the end of natural philosophy', *Social Studies of Science* 16: 387–420

(1989) 'The nebular hypothesis and the science of progress', in Moore (ed.), pp. 131–64

(1991) 'The history and geography of the intellectual world', in Fisch and Schaffer (eds.), pp. 201–31

Schelling, F. (1966) [1803] *On university studies*, trans. E. S. Morgan, with an introduction by N. Guterman, Athens, Ohio

Schiller, F. (1967) [1794] *Letters on the aesthetic education of man*, Oxford

Schipper, F. (1988) 'William Whewell's conception of scientific revolutions', *Studies in History and Philosophy of Science* 19: 43–53

Schneewind, J. B. (1968) 'Whewell's ethics', *American Philosophical Quarterly Monograph* 1, 108–41

(1977) *Sidgwick's ethics and Victorian moral philosophy*, Oxford

Schuster, J. A. and R. R. Yeo (eds.) (1986) *The politics and rhetoric of scientific method: historical studies*, Boston and Dordrecht

Schweber, S. (1981a) 'Scientists as intellectuals: the early Victorians', in

Paradis and Postlewait (eds.), pp. 1–37

(1981b) *Aspects of the life and thought of Sir John Herschel*, 2 vols., New York

Sedgwick, A. (1834) *A discourse on the studies of the university of Cambridge*, 3rd edn., Cambridge

(1850) *A discourse on the studies of the university of Cambridge with additions and a preliminary dissertation*, 5th edn., London and Cambridge

Seward, G. C. (1938) *Die theoretische Philosophie William Whewells und der Kantische Einfluss*, Tübingen

Shapin, S. (1984) 'Brewster and the Edinburgh career in science', in Morrison-Low and Christie (eds.), pp. 17–24

(1990a) 'Mind is its own place: science and solitude in seventeenth-century England', *Science in Context* 4: 191–218

(1990b) 'Science and the public', in Olby, Cantor, Christie, and Hodge (eds.), pp. 990–1,007

(1991) '"A scholar and a gentleman": the problematic identity of the scientific practitioner in early modern England', *History of Science* 29: 279–327

Sharrock, R. (1962) 'The chemist and the poet: Sir Humphry Davy and the preface to Lyrical Ballads', *Notes and Records of the Royal Society of London* 17: 57–76

Shattock, J. (1989) *Politics and reviewers: the Edinburgh and the Quarterly in the early Victorian age*, Leicester

Shattock, J. and M. Wolff (eds.) (1982) *The Victorian periodical press: soundings and samplings*, Leicester

Shine, H. and H. C. Shine (1949) *The Quarterly Review under Gifford: identification of contributors 1809–1824*, Chapel Hill

Sidgwick, H. (1876) 'Philosophy at Cambridge', *Mind* 1: 235–46

Siegfried, R. and Dott, H. (eds.) (1980) *Humphry Davy on geology. The 1805 lectures for the general audience*, Madison

Smiles, S. (1891) *A publisher and his friends. Memoirs and correspondence of the late John Murray, with an account of the origin and progress of the house, 1768–1843*, 2 vols., London

(1894) *Self help*, London

Smith, S. (1810) 'Calumnies against Oxford', *Edinburgh Review* 16: 158–87

Somerville, M. (1834) *On the connexion of the physical sciences*, London

Southey, R. (1829) 'State and prospects of the country', *Quarterly Review* 39: 475–520

Spencer, H. (1854) 'The genesis of science', *British Quarterly Review* 20: 108–62

Staël-Holstein, A. L. (1813) *Germany*, trans. from the French, 3 vols., London

Stair-Douglas, J. (1881) *The life and selections from the correspondence of William Whewell DD.*, London

Stephen, L. (1885–90) 'William Whewell', *Dictionary of National Biography*, 21 vols., Oxford, vol. xx, pp. 1,365–74

Sterling, J. (1848) *Essays and tales*, collected and edited with a memoir of his life, by Julius Charles Hare, 2 vols., London

Stewart, D. (1793) *Outlines of moral philosophy, for the use of students in the University of Edinburgh*, Edinburgh

  (1828) *The philosophy of the active and moral powers of man*, 2 vols., Edinburgh

  (1854–60) *The collected works of Dugald Stewart*, ed. W. Hamilton, 11 vols., Edinburgh

Stocking, G. (1987) *Victorian anthropology*, New York

Sullivan, A. (ed.) (1983) *British literary magazines. The Victorian and Edwardian age, 1837–1913*, Connecticut

Swainson, W. (1834) *Preliminary discourse on the study of natural history*, London

Theerman, P. (1985) 'Unaccustomed role: the scientist as historical biographer – two nineteenth-century portrayals of Newton', *Biography* 8: 145–62

Thomson, T. (1812) *History of the Royal Society from its institution to the end of the eighteenth century*, London

  (1830) *The history of chemistry*, London

Thomson, W. (1868) *The limits of philosophical inquiry*, Edinburgh

Todhunter, I. (1876) *William Whewell, DD. Master of Trinity College Cambridge: an account of his writings with selections from his literary and scientific correspondence*, 2 vols., London

Tulloch, J. (1868) 'The Positive philosophy of M. Auguste Comte', *Edinburgh Review* 127: 303–57

Turner, F. M. (1978) 'The Victorian conflict between science and religion: a professional dimension', *Isis* 69: 356–76

[Ulrici, ?] (1847) 'On Whewell's inductive sciences', *United Services Magazine*, in WP, 266 c. 80[147]

Ward, A. (1974) *Book production, fiction and the German reading public 1740–1800*, Oxford

Weber, M. (1989) *Max Weber's 'science as a vocation'*, ed. P. Lassmann and I. Velody, London

Westfall, R. (1980) *Never at rest. A biography of Isaac Newton*, Cambridge

Wettersten, J. and J. Agassi (1991), 'Whewell's problematic heritage', in Fisch and Schaffer (eds.), pp. 345–69

Whately, R. (1829) 'Logic', *Encyclopaedia Metropolitana* 1: 193–240

Whewell, W. (1819) *An elementary treatise on mechanics*, Cambridge

Whewell, W. (1823) *A treatise on dynamics*, Cambridge

  (1828a) *Statement respectfully offered to the members of the senate*, 9 December, WP, 266 c. 80[49]

  (1828b) *An essay on mineralogical classification and nomenclature*, Cambridge

  (1830a) *Architectural notes on German churches*, new edn, London

  (1830b) 'Mathematical exposition of some doctrines of political economy', *Transactions of the Cambridge Philosophical Society* 3: 191–230

  (1831a) 'Modern science – inductive philosophy', *Quarterly Review* 45: 374–407

  (1831b) 'English universities', *British Critic* 9: 71–90

(1831c) 'Lyell's *Principles of geology'*, *British Critic* 9: 180–206

(1831d) 'Jones – on the distribution of wealth and the sources of taxation', *British Critic* 10: 41–61

(1831e) 'The progress of geology', *Edinburgh New Philosophical Journal* 11: 242–67

(1831f) 'Mathematical exposition of some leading doctrines in Mr. Ricardo's principles of political economy and taxation', *Transactions of the Cambridge Philosophical Society* 4: 155–98

(1832a) 'Lyell's *Principles of geology, volume two'*, *Quarterly Review* 93: 103–32

(1832b) 'Report on the recent progress and present state of mineralogy', *Report of the first and second meetings of the British Association for the Advancement of Science* (1833), London, 322–65

(1832c) *The first principles of mechanics: with historical and practical illustrations,* Cambridge

(1832d) *An introduction to dynamics containing the laws of motion and the first three sections of the Principia,* Cambridge

(1832e) *On the free motion of points and on universal gravitation . . . the first part of a new edition of a treatise on dynamics,* Cambridge

(1833a) 'Address', *Report of the third meeting of the British Association for the Advancement of Science* (1834), London, xi–xxvi

(1833b) 'On the uses of definitions', *Philological Museum* 2, 263–72

(1834a) [1833] *Astronomy and general physics considered with reference to natural theology,* 2nd edn, London

(1834b) 'Mrs Somerville on the connexion of the sciences', *Quarterly Review* 51: 54–68

(1834c) 'Reply to the Edinburgh Review', *British Magazine* 59: 263–68

(1834d) *Remarks on some parts of Mr Thirlwall's letter on the academic admission of dissenters to academical degrees,* Cambridge

(1835a) *Thoughts on the study of mathematics as part of a liberal education,* Cambridge

(1835b) 'Report on the recent progress and present condition of the mathematical theories of electricity, magnetism and heat', *Report on the fifth meeting of the British Association for the Advancement of Science* (1836), London, 1–34

(1836) *Newton and Flamsteed: remarks on an article in number 109 of the Quarterly Review,* Cambridge

(1837a) *History of the inductive sciences, from the earliest to the present time,* 3 vols., London

(1837b) *To the editor of the Medical Gazette,* 11 December, WP, 266 Add. MS. c. 80 149[15]

(1837c) *The mechanical Euclid, containing the elements of mechanics and hydrostatics,* Cambridge

(1837d) *On the foundation of morals: four sermons preached before the university of Cambridge,* Cambridge and London

(1837e) *To the editor of the Edinburgh review,* WP, 266 Add. MS. c. 80

(1838a) *On the principles of English university education*, 2nd edn, London

(1838b) *Address delivered at the anniversary meeting of the Geology Society of London*, London

(1840a) *The philosophy of the inductive sciences, founded upon their history*, 2 vols., London

(1840b) *Remarks on the review of the 'Philosophy of the inductive sciences' in The Athenaeum, no. 672, Sept. 12, 1840*, 22 September 1840, Cambridge, WP, 266 c. 80

(1841a) *Two introductory lectures on moral philosophy*, Cambridge

(1841b) *The mechanics of engineering*, London

(1841c) 'Address', *Report of the eleventh meeting of the British association for the advancement of science* (1842), London, xxvii–xxxv

(1842) *Architectural notes on German churches: with notes written during an architectural tour in Picardy and Normandy*, 3rd edn., London

(1844) 'On the fundamental antithesis of philosophy', *Transactions of the Cambridge Philosophical Society* 7, part 2: 170–81

(1845a) *Of a liberal education in general, and with particular reference to the leading studies of the University of Cambridge*, London

(1845b) *The elements of morality, including polity*, 2 vols., London

(1846a) *Indications of the creator*, 2nd edn, London

(1846b) *Lectures on systematic morality*, London

(1846c) *Newton's Principia: Book 1, sections 1, 11, 111 in the original Latin; with explanatory notes and references*, London

(1847a) *The philosophy of the inductive sciences, founded upon their history*, 2nd edn, 2 vols., London

(1847b) *Verse translations from the German, including Lenore, Schiller's Song of the Bell*, London

(1848a) *Butler's three sermons on human nature*, Cambridge and London

(1848b) 'Second memoir on the fundamental antithesis of philosophy', *Transactions of the Cambridge Philosophical Society* 8: 614–20

(1849a) 'On Mr Macaulay's praise of superficial knowledge', *Fraser's Magazine* 40: 171–5

(1849b) *Of induction*, London

(1851a) *The general bearing of the Great Exhibition on the progress of art and science*, London

(1851b) 'On the transformation of hypotheses in the history of science', in Whewell (1860), pp. 492–503

(1853) *Of the plurality of worlds: an essay*, London

(1854a) *The influence of the history of science upon intellectual education*, London

(1854b) *On the material aids of education*, London

(1854c) *Of the plurality of worlds: an essay*, 3rd edn, London

(1857a) *History of the inductive sciences from the earliest to the present time*, 3rd edn, 3 vols., London

(1857b) 'Spedding's complete edition of the works of Bacon', *Edinburgh Review* 106: 287–322

(1858a) *Novum organon renovatum*, London
(1858b) *The history of scientific ideas*, 2 vols., London
(1859) *Barrow and his academical times*, Cambridge
(1860) *On the philosophy of discovery*, London
(1862) [1852] *Lectures on the history of moral philosophy*, Oxford
(1866) 'Comte and positivism', *Macmillan's Magazine* 13: 353–62
Williams, P. (1991) 'Passing on the torch: Whewell's philosophy and the
  principles of English university education', in Fisch and Schaffer (eds.),
  pp. 117–47
Wilson, D. B. (1974) 'Herschel and Whewell's version of Newtonianism',
  Journal of the History of Ideas 35: 79–97
Winstanley, D. A. (1940) *Early Victorian Cambridge*, Cambridge
  (1947) *Later Victorian Cambridge*, Cambridge
Wiseman, N. (1853) *Essays on various subjects*, 3 vols., London
Wolf, A. (1938) *A history of science, technology and philosophy in the eighteenth
  century*, London
Wolff, M. (1959) 'Victorian reviewers and cultural responsibility', in
  P. Appleton, W. Madden, and M. Wolff (eds.), *Entering an age of crisis*,
  Bloomington, pp. 269–89
Yeo, R. R. (1977) 'Natural theology and the philosophy of knowledge in
  Britain, 1819–1869', unpubl. Ph.D. thesis, University of Sydney
  (1979) 'William Whewell, natural theology and the philosophy of science
  in mid-nineteenth-century Britain', *Annals of Science* 36: 493–512
  (1981) 'Scientific method and the image of science, 1831–1891', in R. M.
  Macleod and P. Collins (eds.), *The parliament of science: the British
  Association for the Advancement of Science, 1831–1981*, Northwood, pp. 65–88
  (1984) 'Science and intellectual authority in mid-nineteenth-century
  Britain: Robert Chambers and *Vestiges of the natural history of creation*',
  Victorian Studies 28: 5–31
  (1985) 'An idol of the marketplace: Baconianism in nineteenth-century
  Britain', *History of Science* 23: 251–98
  (1986a) 'Scientific method and the rhetoric of science in Britain,
  1830–1917', in J. A. Schuster and R. R. Yeo (eds.) (1986), pp. 259–
  297
  (1986b) 'The principle of plenitude and natural theology in nineteenth-
  century Britain', *British Journal for the History of Science* 19: 263–82
  (1987) 'William Whewell on the history of science', *Metascience* 5: 25–40
  (1988) 'Genius, method and morality: images of Newton in Britain,
  1760–1860', *Science in context* 2: 257–84
  (1989) 'Reviewing Herschel's *Discourse*', *Studies in History and Philosophy of
  Science* 20: 541–52
  (1991a) 'William Whewell's philosophy of knowledge and its reception',
  in Fisch and Schaffer (eds.), pp. 175–99
  (1991b) 'Reading encyclopaedias: science and the organisation of knowledge in British dictionaries of arts and sciences', *Isis* 82: 24–49

Young, G. M. (1936) *Portrait of an age: Victorian England*, London
Young, R. M. (1985) *Darwin's metaphor: nature's place in Victorian culture*, New York
Young, T. (1810) 'Mémoires d'Arcueil', *Quarterly Review* 3: 462–81

Index

Abercrombie, John, 188
Albert, Prince Consort, 231
Alter, Peter, 32
anonymity, in reviews of science, 82–3
audience, for reviews of science, 108–9, 114

Babbage, Charles, 24, 33–5, 46, 91, 125,
    212, 227–8
  dispute with Whewell, 123–4
  on unity of science, 235
Bacon, Francis, 10, 12, 63, 166, 214
  and Comte, 247
  Coleridge on, 183
  his doctrine of idols, 135–6
  Whewell on, 10–11, 247
Baconian method, 96–7, 106–7, 128, 162,
    240, 250
  relation to empiricism, 177, 182–4
  versus solitude, 135–7
Baily, Francis, 129–30
  dispute with Whewell, 133–4
Barrow, Isaac, 172
Becher, Harvey, 218–19
Bentham, Jeremy, 184
Berkeley, George, 57
Bichat, Xavier, 152
biography of scientists, 6, 8, 116–18
biology, 254
  Whewell vs Powell, on 242–3
Biot, J.B., 18, 142
Birks, Thomas, 198
Blair, Alexander, 60
Boadicea, Whewell's poem on, 57
Bowden, William, 126–7
Boyle, Robert, 136
Brewster, David, 21–2, 35, 45, 72, 86, 114
  criticism of Whewell, 141, 164, 225,
    233–4
  on Baconian method, 164
  on genius, 164
  on Newton, 139–42
  on research, 91–2

on reviews of science, 80–1, 85n
on status of science, 86
on technology, 225
Bridgewater Treatises, 118
British Critic, 90–2
British Magazine, 90
British Association for the Advancement of
  Science, 32, 72, 223, 231
Brooke, John, 31, 248
Brougham, Henry, 38, 45, 80–1, 83, 119,
  127, 136
Brown, Thomas, 183
Buckland, William, 32
Bulwer-Lytton, Edward, 37
Burckhardt, Jacob, 158n
Burke, Edmund, 168
Burrow, J.W., 168–9
Butler, Samuel, 200–1
Butterfield, H., 168
Butts, Robert, 243–4

Cabinet Cyclopaedia, 86, 88
Calderwood, Henry, 246
Cannon, Susan F. (W.F.), 29–32, 44, 83
  on 'Truth Complex', 30, 38
Cantor, G.N., 147
Carlyle, Thomas, 3, 140, 148
Chalmers, Thomas, 188
Chambers, Robert, 113, 232–3
  and specialization, 113–14
character, of scientists, 116–17, 119–20
classics, Whewell on, 215–16
classification of sciences, 49, 233
  see also unity of science
Cockburn, William, 79, 124, 126–7
Cohen, I.B., 165
Coleridge, Samuel Taylor, 3, 51, 58–9, 60,
  110, 183
Coleridge, Samuel Taylor (cont.)
  idea of clerisy, 44, 52, 202, 222
  Whewell on, 66–7
communication of science, 35–8, 86–7, 113–14

Comte, Auguste, 49–52, 148, 159, 233–4
Condillac, Etienne Bonnot, de, 182
Copleston, Edward, 78, 84
Corsi, Pietro, 23, 31
Cuvier, Georges, 10, 36, 56, 100

Dalton, John, 97
Darwin, Charles, 4, 30–1, 39, 56, 235
Davy, Humphry, 5, 30, 65, 124
De Morgan, Augustus, 10, 13, 21, 23, 86,
    125, 148, 189, 211n,
    on Newton, 142
definitions, 107
DeQuincey, Thomas, 60, 71
Descartes, René, Whewell's opinion of, 172
design argument, 118–19, 188–9, 196, 248
disciplines, 106, 239–41, 251, 253
discovery, and scientific reputations, 6, 8,
    52–55
Duhem, Pierre, 158n

Eagleton, Terry, 43
Edgeworth, Maria, 141
Edinburgh Review, 42–3, 78, 81–3, 87
Elkana, Y., 147, 157
empiricism, 61, 177–8, 180, 182–6, 190, 200
    see also Baconian method, idealism, Locke
Encyclopaedia Britannica, 14, 24, 58
    and histories of science, 150
Encyclopaedia Metropolitana, 47, 60, 87
encyclopaedias, and omniscience, 58–60
    and specialization of science, 33
epistemology and morals, 187–90

Faraday, Michael, 5–6, 21, 34, 231, 242
Farrar, Adam, 248
Feyerabend, Paul, 4
First Reform Bill, 37, 46
Fisch, Menachem, 7, 61–2, 68, 146–7, 190,
    219
Flamsteed, John, 129–34
    Whewell on, 131–2, 134
Forbes, Duncan, 168
Forbes, James David, 24, 34, 38, 112, 129,
    132, 137, 173
Fox, Caroline, 33, 56
fundamental antithesis, in Whewell's
    philosophy, 11–13, 190, 241, 253
fundamental ideas, 12–13, 189, 197, 161,
    200, 215–7, 220, 239–40, 242–3, 249,
    252–3

Galloway, Thomas, 131
Galton, Francis, 5, 116
Gascoigne, John, 32
Geological Society, Whewell's Presidency
    of, 73

geology, 33, 96, 99–102
    see also Lyell
Goethe, Johann Wolfgang, 10
Gothic architecture, and Whewell's
    historical views, 154–5
Great Exhibition, 224–5

Habermas, Jürgen, 40–1, 43–4, 46, 77
Hahn, Roger, 35
Halliwell, J.O., 149
Hamilton, Sir William, 43, 179, 212,
    218–19, 245
Hamilton, William Rowan, 23, 69–70, 123,
    129, 137
Harcourt, William Vernon, 51, 111, 124,
    128
Hare, Augustus, 152, 155
Hare, Julius Charles, 7, 19, 66, 123n,
    202–5
Herschel, John, 7, 19, 21–2, 29, 47, 61, 129
    and solitude, 141
    his *Discourse*, 28, 39, 72, 92–8, 119–20
    on epistemology, 179–80, 185–89
    on history of science, 148, 156
    on science and society, 229
Heyck, T.W., 41
hierarchy of sciences, 106–8
Hilton, Boyd, 194
historical consciousness, 149
history of science, as way of defining
    science, 145–9, 150–2
    Whewell's account, 145–76
Hodge, M.J.S., 15
Houghton, Walter, 77
Humboldt, Alexander von, 229
Humboldt, Wilhelm von, 41
Hume, David, 149
Husserl, Edmund, 4
Huxley, Thomas Henry, 5, 39, 243
hypothesis, method of, 63, 96–7

idealism, 4, 12–13, 177–80, 185, 204–5
induction, in logic, 12–13, 63, 93
inductive method, 12–13, 62–3, 92–4, 96–8,
    104, 120–4, 160–1
    see also Baconian method, method
institutions, as locus of scientific authority,
    124–6
    Tractarian critique of, 126–7
intellectual authority, 37, 86, 254

Jeffrey, Francis, 43, 45–6
Jones, Richard, 7, 21, 23, 189
    on inductive philosophy, 62, 104
    on morals, 105
    on political economy, 102–6, 195–7

Kant, Immanuel, 13–14, 47, 192, 249
Kepler, Johanes, 10
  as genius, 120–1, 162
Knightbridge Chair, 189
Kuhn, Thomas S., 35

Lakatos, Imre, 4
Laplace, Pierre, 116
Lardner, Dionysius, 86
Laudan, Larry, 50
Laudan, Rachel, 151n
Lavoisier, Antoine, 107, 166
Lewis, G.C., 173
Liberal Anglicanism, and view of history,
  155, 174
liberal education, 91, 209, 212–18
  mathematics in, 218–22
  science in, 210–11, 215–22
Locke, John, his theory of knowledge, 13,
  180, 226
  Sedgwick on, 176, 180–2
  Whewell on, 191–3
Lockhart, John, 83–4
Lubbock, John, 54
Lyell, Charles, 6–7, 24, 37, 72–3, 85, 146,
  226
  as geological theorist, 99–102
  on history of science, 159
  on scientific publishing, 80
  on universities, 210–11

Macaulay, T.B., 56, 58–9, 158n, 168–9,
  172, 226–7
McCosh, James, 246
McCulloch, J.R., 80
Mackintosh, James, 198–9
MacLeod, R.M., 32
MacMillan's Magazine, 234
Malcolm, John, 17
Manning, Henry, 247
Mansel, Henry, 245–7
Marshall, Cordelia, 15–16
Marshall, James Garth, 59, 167
Martineau, Harriet, 233–5
Martineau, James, 245–6, 249
mathematics, Whewell on, 218–21
Maurice, F.D., 20
Maxwell, James Clerk, 5, 110, 231
medieval period, Whewell's attitude to, 158
mental habits, and natural theology, 120–4
Merton, Robert, 117
Merz, John T., 28, 49, 117, 140
metascience, 8–9
  as a role, 11, 50–2, 56, 63–5, 70–4,
  146–8, 252, 254–5

contemporary attitude to, 50–2, 72–4,
  233–4
method
  and moral character, 118–24
  and unity of science, 235–6, 240
  as accessible, 161–2
  vs genius, 140–1, 162–5
  vs personality, 117, 136
Michelet, Jules, 78
Mill, John Stuart, 4, 37, 42, 65, 98, 204–5,
  243, 250
  on empiricism and intuitionism, 178,
  184–5, 236
  on Whewell, 178, 184–5, 205, 236, 255
mineralogy, Whewell's work in, 53
Moll, Gerrit, 79
Moore, James, 56
moral philosophy, 176, 180–2, 192–3,
  198–200, 236–9
  see also utilitarianism
moral science, analogies with physical
  science, 236–9
Moral Sciences Tripos, Cambridge, 245
Morell, John Daniel, 246, 248
Morland, George, 57, 191
Morrell, J.B., 32, 117, 128
Munby, A.N.L., 149
Murchison, Roderick, 20, 52, 125
Murray, John, 80, 83
Myer, Frederick, 113, 169, 237–9

Napier, Macvey, 43–4, 113, 150
natural philosophy, as generic term, 24, 33
natural theology, 29–32
  and epistemology, 187–9
  and geology, 102
  and method, 120–4, 187–9
natural history, 33, 241
'nescience', 246, 250
Newman, Francis, 209, 222
Newman, John Henry, 30, 79
  on moral effects of science, 119
Newton, Sir Isaac, 10, 130–4, 220–1
  as genius, 140–3
  as moral exemplar, 139, 144
  vs Flamsteed, 130–4
Niebuhr, Barthold, 148, 155

omniscience, 57–60
original research, 91, 212–13
Outram, Dorinda, 36, 56, 135–6
Owen, Richard, 59, 83, 173

Paley, William, 180–2, 235, 238
  on utilitarian ethics, 181

Passmore, John, 178
Pasteur, Louis, 5
Pattison, Mark, 245
Peacock, George, 147
Peacock, Thomas, 58, 66, 157
Peel, Sir Robert, 19, 32
periodical reviews, 38, 42, 77–90
  see also anonymity; scientific reviewing
permanent sciences, Whewell's definition
  of, 215–16, 220–4
physiology, 173, 173n, 243, 251
Playfair, John, 36–7, 81, 150
political economy, 102–6
  and inductive method, 103–4
  and morals, 105–6, 194–6
  Ricardian theory, 105–6
  Whewell on, 102–6, 194–7
  see also Jones
Popper, Karl, 4
Positivism, 243–7, 249
Powell, Baden, 21–3, 64, 249–50
  on epistemology, 186
  on history of science, 159
  on Newton, 143
  on principle of uniformity, 235–6
  on science education, 79, 86, 91
  on unity of science, 235–6, 241–3
Preyer, Robert, 66
Priestley, Joseph, 161
progress of science, 153–60
  and novelty, 68–9, 167, 170
  place of accident in, 164
  Whewell's three stages of, 155–6
progressive sciences, Whewell's definition
  of, 216–17, 220–4
public sphere, 25, 40–4
  and science, 44–7, 112–14

Quarterly Journal of Education, 86
Quarterly Review, 42, 81–3, 85

Reid, Thomas, 57, 93
Remusat, Charles, 247
Rennell, Thomas, 188
revolutions in science, Whewell on, 166–72
Ricardo, David, 103, 105
Richards, Joan, 179
Rigaud, Stephen, 130
Roget, Peter Mark, 51
Romanticism, and cultural criticism, 70–1
  and science, 65–71
  and Newton, 139–40
Romilly, Joseph, 18
Rose, Hugh James, 24, 67, 88, 90, 167, 192
  critical of science, 68–9, 123, 202–3

Rowley, Joseph, 57
Royal Society of London, 33, 46
Ruse, Michael, 53

Saint Simon, Henri de, 197, 214
Sarton, George, 251
Schaffer, Simon, 7, 70, 89
Schelling, Friedrich, 213
Schiller, Friedrich, 10, 70–1
Schipper, Frits, 166
Schlegel, August Wilhelm, 71
science, 10, 32–3, 106
  and poetry, 65–70
  as vocation, 34–6
  language of, 6, 39–40, 110–11
scientific books, 80–2
scientific reviewing
  attitude to, 85–7
  style in, 83–5
  see also periodical reviews; scientific books
scientist, as generic term, 110–11
Sedgwick, Adam, 7, 24, 73–4, 176
  his *Discourse*, 176, 180–2, 193
Senior, Nassau, 105–6, 195
Sewell, William, 119
Shapin, Steven, 135
Shelley, P.B., 66
Sidgwick, Henry, 18, 245
Smiles, Samuel, 85, 162n
Smith, William, 100
Smith, Sydney, 6, 56
solitude, 136–8, 141
Somerville, Mary, 80–1, 109–12
Southey, Robert, 37
specialization, 7, 33–5
  and periodicals, 46–7, 84
  as problem for Whewell, 11, 147, 172–3,
    251–4
  vs synthetic views of science, 113–14
  see also unity of science
Spencer, Herbert, 49–50, 251
Sprat, Thomas, 135
Staël-Holstein, Anne Louise Germaine de,
  71, 192
Stephen, Leslie, 18, 24, 52
Sterling, John, 33, 77, 201, 204
Stewart, Dugald, 10, 58, 63–4, 136
  and classification of sciences, 58
  on geometry, 179
  on mathematicians, 116, 119n
Strauss, D.F., 248
Sumner, J.B., 194

technology, as distinct from science, 225–30
Thackray, Arnold, 32, 128

theorists, their role and character, 127–9,
132, 135, 138–9, 144
Thirlwall, Connop, 213
Thomson, Thomas, 151
Thomson, William, 245
tides, 54–5
Tocqueville, Alexis de, 37n
Todhunter, Isaac, 7, 62, 231
Tractarians, 119
on science, 126–7
Trinity College Cambridge, 3, 6, 15–16,
103
Tulloch, John, 246, 249
Turner, F.M., 29

unity of science, 110–11, 231–3
in Herschel's *Discourse*, 94–6
limits to, 239–41
Somerville on, 110
Whewell vs Powell on, 241–3
universities, criticisms of, 86, 90–2, 210
see also liberal education
utilitarianism, 180–2, 186–7, 196, 199–203,
223–4, 232, 237

Weber, Max, 4, 34
Werner, Abraham, 100
Westminster Review, 42, 82–3, 90, 204, 214
Whately, Richard, 10, 93, 105–6, 122, 194
Whewell, William
academic positions, 15
early life, 16–19, 57–8
marriage, 16–17
Master of Trinity College, 18–21
reputation, 4–8, 52
scientific work, 53–5

textbooks, 55, 61
and D. Brewster, 91–2, 141, 164–5
and W.R. Hamilton, 69–70
and J.S. Mill, 4, 177–8, 185–6, 204,
250–1, 255
on Bacon, 10–11, 96–7, 152, 154, 247,
254
on Comte, 244, 246, 250
on mathematics, 179, 218–21
on morals and induction, 198–200
on natural theology, 118–19, 194–5,
247–8
on political economy and epistemology,
196–8
on reviewing, 87–90
on Romanticism, 66–71
on technology, 224–9
reviews Herschel's *Preliminary Discourse*,
92–9
reviews Jones's political economy, 102–6
reviews Lyell's *Principles of geology*, 99–102
reviews Somerville's *Connexion*, 109–12
Whig history, 150–1, 168–9, 174
Wilberforce, Samuel, 254
Williams, Perry, 211
Wiseman, Nicholas, 184
Wolff, Michael, 78
Wollaston, William, 124
women in science, 35, 70, 111–12
Wordsworth, Christopher, 19, 67
Wordsworth, William, 65–7, 70

Young, G.M., 77
Young, R.M., 31–2, 44–5
on Victorian periodicals, 38–9, 46
Young, Thomas, 29, 79, 112, 124

Ideas in Context

Edited by Quentin Skinner (general editor), Lorraine Daston, Wolf Lepenies, Richard Rorty and J. B. Schneewind

1. RICHARD RORTY, J. B. SCHNEEWIND and QUENTIN SKINNER (eds.)
Philosophy in history
Essays in the historiography of philosophy *
2. J. G. A. POCOCK
Virtue, commerce and history
Essays on political thought and history, chiefly in the eighteenth century *
3. M. M. GOLDSMITH
Private vices, public benefits
Bernard Mandeville's social and political thought
4. ANTHONY PAGDEN (ed.)
The languages of political theory in early modern Europe *
5. DAVID SUMMERS
The judgment of sense
Renaissance nationalism and the rise of aesthetics *
6. LAURENCE DICKEY
Hegel: religion, economics and the politics of spirit, 1770–1807 *
7. MARGO TODD
Christian humanism and the Puritan social order
8. LYNN SUMIDA JOY
Gassendi the atomist
Advocate of history in an age of science
9. EDMUND LEITES (ed.)
Conscience and casuistry in early modern Europe
10. WOLF LEPENIES
Between literature and science: the rise of sociology *
11. TERENCE BALL, JAMES FARR and RUSSELL L. HANSON (eds.)
Political innovation and conceptual change *
12. GERD GIGERENZER et al.
The empire of chance
How probability changed science and everyday life *

13. PETER NOVICK
That noble dream
The 'objectivity question' and the American historical profession *
14. DAVID LIEBERMAN
The province of legislation determined
Legal theory in eighteenth-century Britain
15. DANIEL PICK
Faces of degeneration
A European disorder, c.1848–c.1918
16. KEITH BAKER
Approaching the French Revolution
Essays on French political culture in the eighteenth century *
17. IAN HACKING
The taming of chance *
18. GISELA BOCK, QUENTIN SKINNER and MAURIZIO VIROLI (eds.)
Machiavelli and republicanism *
19. DOROTHY ROSS
The origins of American social science *
20. KLAUS CHRISTIAN KÖHNKE
The rise of Neo-Kantianism
German Academic Philosophy between Idealism and Positivism
21. IAN MACLEAN
Interpretation and meaning in the Renaissance
The Case of Law
22. MAURIZIO VIROLI
From politics to reason of state
The Acquisition and Transformation of the Language of Politics 1250–1600
23. MARTIN VAN GELDEREN
The political thought of the Dutch revolt 1555–1590
24. NICHOLAS PHILLIPSON and QUENTIN SKINNER (eds.)
Political discourse in early modern Britain
25. JAMES TULLY
An approach to political philosophy: Locke in contexts *
26. RICHARD TUCK
Philosophy and government 1572–1651 *
27. RICHARD YEO
Defining science
William Whewell, natural knowledge and public debate in early Victorian Britain

Titles marked with an asterisk are also available in paperback